AWS D10.9-80

# Specification for Qualification of Welding Procedures and Welders for Piping and Tubing

Superseding AWS D10.9-69

Prepared by
AWS Committee on Piping and Tubing

Under the Direction of
AWS Technical Activities Committee

Approved by
AWS Board of Directors, October 6, 1980

**AMERICAN WELDING SOCIETY, INC.**
550 North LeJeune Rd., Miami, FL 33126

**Statement on Use of AWS Standards**

All standards (codes, specifications, recommended practices, methods, classifications, and guides) of the American Welding Soc are voluntary consensus standards that have been developed in accordance with the rules of the American National Standa Institute. When AWS standards are either incorporated in, or made part of, documents that are included in federal or state l and regulations, or the regulations of other governmental bodies, their provisions carry the full legal authority of the statute. such cases, any changes in those AWS standards must be approved by the governmental body having statutory jurisdiction be they can become a part of those laws and regulations. In all cases, these standards carry the full legal authority of the cont or other document that invokes the AWS standards. Where this contractual relationship exists, changes in or deviations fr requirements of an AWS standard must be by agreement between the contracting parties.

International Standard Book Number: 0-87171-21

American Welding Society, 550 N.W. LeJeune Road, P. O. Box 351040, Miami, Florida 33



Printed in the United States of Ame
Reprinted: February 1

Note: The primary purpose of AWS is to serve and benefit its members. To this end, AWS provides a forum for the excha consideration, and discussion of ideas and proposals that are relevant to the welding industry and the consensus of which fo the basis for these standards. By providing such a forum, AWS does not assume any duties to which a user of these stand may be required to adhere. By publishing this standard, the American Welding Society does not insure anyone using information it contains against any liability arising from that use. Publication of a standard by the American Welding Society d not carry with it any right to make, use, or sell any patented items. Users of the information in this standard should make independent investigation of the validity of that information for their particular use and the patent status of any item referre herein.

With regard to technical inquiries made concerning AWS standards, oral opinions on AWS standards may be rendered. Howe such opinions represent only the personal opinions of the particular individuals giving them. These individuals do not speak behalf of AWS, nor do these oral opinions constitute official or unofficial opinions or interpretations of AWS. In addition, opinions are informal and should not be used as a substitute for an official interpretation.

This standard is subject to revision at any time by the AWS Committee on Piping and Tubing. It must be reviewed every years and if not revised, it must be either reapproved or withdrawn. Comments (recommendations, additions, or deletions) any pertinent data that may be of use in improving this standard are requested and should be addressed to AWS Headquar Such comments will receive careful consideration by the AWS Committee on Piping and Tubing and the author of the comm will be informed of the Committee's response to the comments. Guests are invited to attend all meetings of the AWS Commi on Piping and Tubing to express their comments verbally. Procedures for appeal of an adverse decision concerning all comments are provided in the Rules of Operation of the Technical Activities Committee. A copy of these Rules can be obta from the American Welding Society, 550 N.W. LeJeune Road, P. O. Box 351040, Miami, Florida 33135.

# Table of Contents

# Personnel

**Committee on Piping and Tubing**

| | |
|---|---|
| *R.K. Sager, Chairman* | Consultant |
| *R.R. Wright, 1st Vice-Chairman* | Moody International, Inc. |
| *R. Giambelluca, 2nd Vice-Chairman* | C.F. Braun and Company |
| *G. Hallstrom, Jr., Secretary* | American Welding Society |
| *W.L. Ballis* | Columbia Gas Distribution Company |
| *M.T. Collier* | Consultant |
| *K.L. Crooks* | Armco, Inc. |
| *G. Curbow* | Fuchs Machinery |
| *H.W. Ebert* | Exxon Research and Engineering |
| *T.A. Ferguson* | CRC-Automatic Welding |
| *J.E. Fisher* | Speri Associates |
| *R.S. Green* | National Certified Pipe Welding Bureau |
| *R.B. Gwin* | H.C. Price Company |
| *E.A. Harwart* | NWI International |
| *G.K. Hickox* | Consultant |
| *J.E. Hinkel* | Lincoln Electric Company |
| *A.N. Kugler** | Consulting Engineer |
| *R.J. Landrum** | E.I. duPont de Nemours |
| *J.R. McGuffey* | Oak Ridge National Laboratory |
| *J.W. Moeller** | The Ralph M. Parsons Company |
| *L.E. Pardi* | Daniel Construction Company |
| *M.D. Randall* | CRC-Automatic Welding |
| *P.C. Shepard* | Consultant |
| *E.G. Shifrin* | Detroit Edison Company |
| *G.K. Sosnin* | Consultant |
| *H.A. Sosnin* | Consultant |
| *W.J. Sperko* | Richmond Engineering |
| *D.W. Storck* | Becon Construction Company, Inc. |
| *J.C. Thompson* | The Ralph M. Parson Company |
| *D.R. VanBuren* | The East Ohio Gas Company |

**Subcommittee on Qualification**

| | |
|---|---|
| *P.C. Shepard, Chairman* | Consultant |
| *G. Hallstrom, Jr., Secretary* | American Welding Society |
| *H.W. Ebert* | Exxon Research and Engineering |
| *R.B. Gwin* | H.C. Price Company |
| *J.E. Hinkel* | Lincoln Electric Company |
| *A.N. Kugler** | Consulting Engineer |
| *R.J. Landrum** | Consultant |
| *J.R. McGuffey* | Oak Ridge National Laboratory |
| *R.K. Sager* | Consultant |
| *E.G. Shifrin* | Detroit Edison Company |
| *G.K. Sosnin* | Consultant |
| *H.A. Sosnin* | Consultant |

*Advisory Member

# Foreword

This specification was prepared by the American Welding Society Committee on Piping and Tubing as a service to the piping industry. This document sets forth the requirements that must be met to qualify welding procedures and welders for joining pipe, tube, and related components.

The Committee believes that the development and application of proven or qualified welding procedures and the qualification of welders to these procedures play a significant role in the making of reproducible, sound welds.

It is recognized that the contractor must provide competent supervision, reliable equipment, and a quality control program during production welding. This specification covers only the qualification of welding procedures and welders; the purchaser shall specify the inspection required during fabrication as part of his contract.

The qualification tests take into account such factors as the diameter of the product, as well as the wall thickness, the positions involved in pipe welding, and the fact that, in most cases, all welding and inspection must be performed from the face side of the weld.

Eight positions related to pipe welding have been set forth, and the nondestructive and destructive tests of the sample welds are tailored to pipe and tube welds.

It should be recognized that there are differences between this document and other codes and standards. Contractors may revise existing welding procedures and welder qualifications to conform with this specification by performing any additional tests required to comply with this document.

# Introduction

This specification provides two distinct levels of quality for welding procedure and welder qualification for joining pipe, tube, and related components. The two levels are called Acceptance Requirements 1 and 3 and are hereafter abbreviated AR-1 and AR-3, respectively. Qualification to the AR-1 level provides a higher level of confidence in welding procedures and welders than does qualification to the AR-3 level. The difference between the two levels lies in the type of tests and the acceptance requirements for results that are specified for sample test welds.

The specification is divided into five sections, plus appendices. Section 1 covers general provisions. Section 2 describes welding procedure qualification for groove welds for the AR-1 level. Section 3 describes welder qualification for groove welds for the AR-1 level. Section 4 covers welding procedure and welder qualification for groove welds for the AR-3 quality level. Section 5 covers welding procedure and welder qualification for the welding of fillet welds only.

Appendix A includes information useful in applying this document. Specifically, it is an attempt at explaining the purpose and mechanics of welding procedure and welder qualification.

This specification sets forth two series of standardized tests and minimum acceptable test results for qualification of welding procedures and welders. The AR-3 level series is directed at noncritical piping systems and is dicusssed at the end of the Appendix. The AR-1 level test series is fairly typical of other national codes and standards for welding procedure and welder qualification, except that it is not directed at any specific application. It may be used as a guide for those developing welding requirements for new applications of piping systems. It may also assist in the understanding of the welding qualification requirements of other standards.

This first revision incorporates major changes and improvements over the 1969 edition to facilitate wider application of the specification. The 1969 edition contained three levels of qualification called Acceptance Requirements 1, 2, and 3, abbreviated as AR-1, AR-2, and AR-3. In this revised edition, the AR-1 and AR-2 Acceptance Requirements have been combined and modified into one level of qualification called Acceptance Requirement 1, abbreviated AR-1. This revised AR-1 level and a revised AR-3 level of Acceptance Requirements comprise the present specification.

A major revision to the specification allows the exclusive use of nondestructive examination for welder qualification without the performance of additional destructive testing if certain criteria are met. Material tables have been updated and regrouped to agree with other standards.

The format has been rearranged to make each section easier to use.

# Specification for Qualification of Welding Procedures and Welders for Piping and Tubing

Issued, 1969; Revised, 1980

## 1. General Provisions

### 1.1 Scope

**1.1.1** This specification sets forth the requirements for qualifying welding procedures and welders[1] for joining ferrous and nonferrous metals and applicable combinations of these materials using any of the processes applicable to tubular products. It covers groove and fillet welds but excludes longitudinal seams involved in pipe and tube manufacture.

**1.1.2** By referencing this specification and specifying the desired level of Acceptance Requirements, any organization may make it the governing document for qualifying welding procedures and welders for the welding of piping and tubing.

**1.1.3** Provision is made in this specification for qualifying welding procedures and welders for both groove and fillet welds. Qualification with groove welds qualifies for welding both groove and fillet welds; whereas, qualification with fillet welds qualifies only for the welding of fillet welds.

**1.1.4** Welding procedures and welders previously qualified to the AR-1 or AR-2 levels of the 1969 edition of this specification are now considered as qualified to the AR-1 level of the present edition of this specification. Welding procedures and welders previously qualified to the AR-3 level or for fillet welds only, in accordance with the 1969 edition, are also considered as qualified to the AR-3 level of the present edition of the specification or for fillet welds only.

### 1.2 Levels of Qualification

**1.2.1** Each of the two levels of welding procedure qualification has a corresponding level of welder qualification. Qualification to the AR-1 level automatically qualifies the procedure and the welder for both the AR-1 and AR-3 levels. Qualification of a welding procedure and a welder to the AR-3 level qualifies only for AR-3.

**1.2.2** Generally, the welder is qualified to the same level as the welding procedure. However, the contractor may use a welding procedure of a different level than that for which the welder is being qualified within the following restrictions:

(1) An AR-1 welding procedure shall be used to qualify a welder for the AR-1 level.

(2) An AR-1 or AR-3 welding procedure may be used to qualify a welder for the AR-3 level.

**1.2.3** When selecting one of the two levels, the tests and Acceptance Requirements specified for each level should be carefully reviewed. Consideration should be given to the service conditions, particularly to the consequences if the system fails to contain the liquid or gas. The level selected should be compatible with such factors as refinement of the design and rigorousness of the inspection and testing procedures specified. The object is to provide the desired degree of confidence in the integrity of the system.

**1.2.4** Typical applications are as follows:

**1.2.4.1 AR-1.** This level applies to systems where a high degree of weld quality is required. Typical applications are lines in nuclear, chemical, cryogenic, gas, or steam systems.

**1.2.4.2 AR-3.** This level applies to systems where an average degree of weld quality is required. Applications are lines such as those for low-pressure heating, air conditioning, sanitary water, and some gases or chemicals.

### 1.3 Responsibility of the Contractor

**1.3.1 Qualification Testing.** Prior to production welding, the contractor shall qualify both his welding procedures and welders in accordance with this specification. Appendix A discusses the purpose and mechanics of welding procedure and welder qualification testing.

1. The term "welder(s)" as used in this document includes both welders and welding operators.

**1.3.2 Welding Procedure Specification.** To qualify a welding procedure, the contractor shall prepare a tentative written welding procedure specification. Sample Form A (see Appendix F) is an example of an acceptable format. The contractor should review paragraph 2.3 to ascertain that the proposed welding procedure specification sets forth all of the controlled welding variables applicable to it. He shall prove the adequacy of this welding procedure by making sample welds, subjecting these welds to the tests, and complying with the Acceptance Requirements set forth herein. Welding procedure qualification test records shall be kept to document the welding variables and record the test results. Sample Form B (see Appendix F) illustrates the items that should be recorded. If the sample welds satisfy the test requirements, a written welding procedure specification shall be prepared, similar to Form A, covering all welding variables significant to the production of sound welds. The welder who makes satisfactory sample welds for the procedure qualification shall be considered qualified to that welding procedure without further testing.

**1.3.3 Duration of Welding Procedure Specification.** The welding procedure shall remain in effect indefinitely, unless one or more of the welding variables undergoes a change greater than that allowed by the limits controlling the qualification (see paragraph 2.3).

**1.3.4 Information for Welders.** The contractor shall make a copy of the welding procedure available to the welder prior to his welder qualification test.

**1.3.5 Welder Qualification.** To qualify a welder, the contractor shall have the welder prepare sample welds using a qualified welding procedure. The contractor shall then subject the welds to tests to verify compliance with the Acceptance Requirements prescribed by this specification. The contractor shall maintain a record of each welder's qualification test, similar in format to Sample Form C (Appendix F). This qualification record shall identify all of the controlled welding variables and contain information on the results of all destructive and nondestructive tests performed on the welder's sample welds.

**1.3.6 Welder Identification.** Each welder qualified shall be assigned an identification number, letter, or symbol that shall be used to identify all of his work.

**1.3.7 Welder Qualification Retest.** A welder whose sample welds fail to meet the Acceptance Requirements of this specification may be retested, at the contractor's discretion, after he has had further training. The retest shall consist of one sample weld in each position in which the welder failed in his previous test.

**1.3.8 Welder Requalification.** Requalification of a welder may be required whenever there is reason to question his ability to make welds that meet the requirements of this specification or when he has not welded for some period of time agreed upon between the purchaser and the contractor.

**1.3.9 Previously Qualified Welding Procedures and Welders**

**1.3.9.1** Welding procedures and welders previously qualified to the AR-1 or AR-2 levels of the 1969 edition of AWS D10.9 are now considered qualified to the AR-1 level of the present edition. Welding procedures and welders previously qualified to the AR-3 level, or for fillet welds only, are also considered as similarly qualified to the present edition. In addition, welding procedures and welders previously qualified in accordance with another standard may also be considered qualified to this specification if the results of such qualification tests meet all of the Acceptance Requirements set forth herein for the level of qualification selected.

**1.3.9.2** If tests required by this specification have not previously been performed, the contractor may update the welding procedure and welder qualifications by successful completion of only those additional tests needed to comply with the Acceptance Requirements of this specification. If radiographic, liquid penetrant, or magnetic particle examination are the only additional tests needed for compliance with this specification, they may be made on the next production welds.

**1.3.9.3** Bend tests, successfully completed to the Acceptance Requirements of this specification but using a smaller radius than specified by this specification for the material in question, are considered to meet the requirements of this specification.

**1.4 Test Positions**

**1.4.1** The welding of piping and tubing systems often involves welding in several different positions. The proper conditions for welding in each position must be established as a part of the procedure qualification. To facilitate testing, standard positions have been defined for both groove and fillet welds.

**1.4.2** For groove welds in pipe, the four standard positions are the 1G, 2G, 5G, and 6G positions, as shown in Fig. 1.4.2. Qualification in one groove weld position qualifies for welding in other positions as follows:

(1) Qualification in the 1G position qualifies only for that position.

(2) Qualification in the 2G position qualifies for welding in the 1G and 2G positions only.

(3) Qualification in the 5G position qualifies for welding in the 1G and 5G positions only.

(4) Qualification in both the 2G and 5G positions qualifies for welding in any position.

(5) Qualification in the 6G position qualifies for welding in any position because it includes elements of all positions.

(6) Qualification in a position other than the four standard positions described above is valid only for that position (plus or minus 15 degrees).

**1.4.3** For fillet welds in pipe, the four standard positions are the 1F, 2F, 4F, and 5F positions as shown in Fig. 1.4.3 Qualification in any of the fixed G positions, using sample groove welds in pipe, qualifies

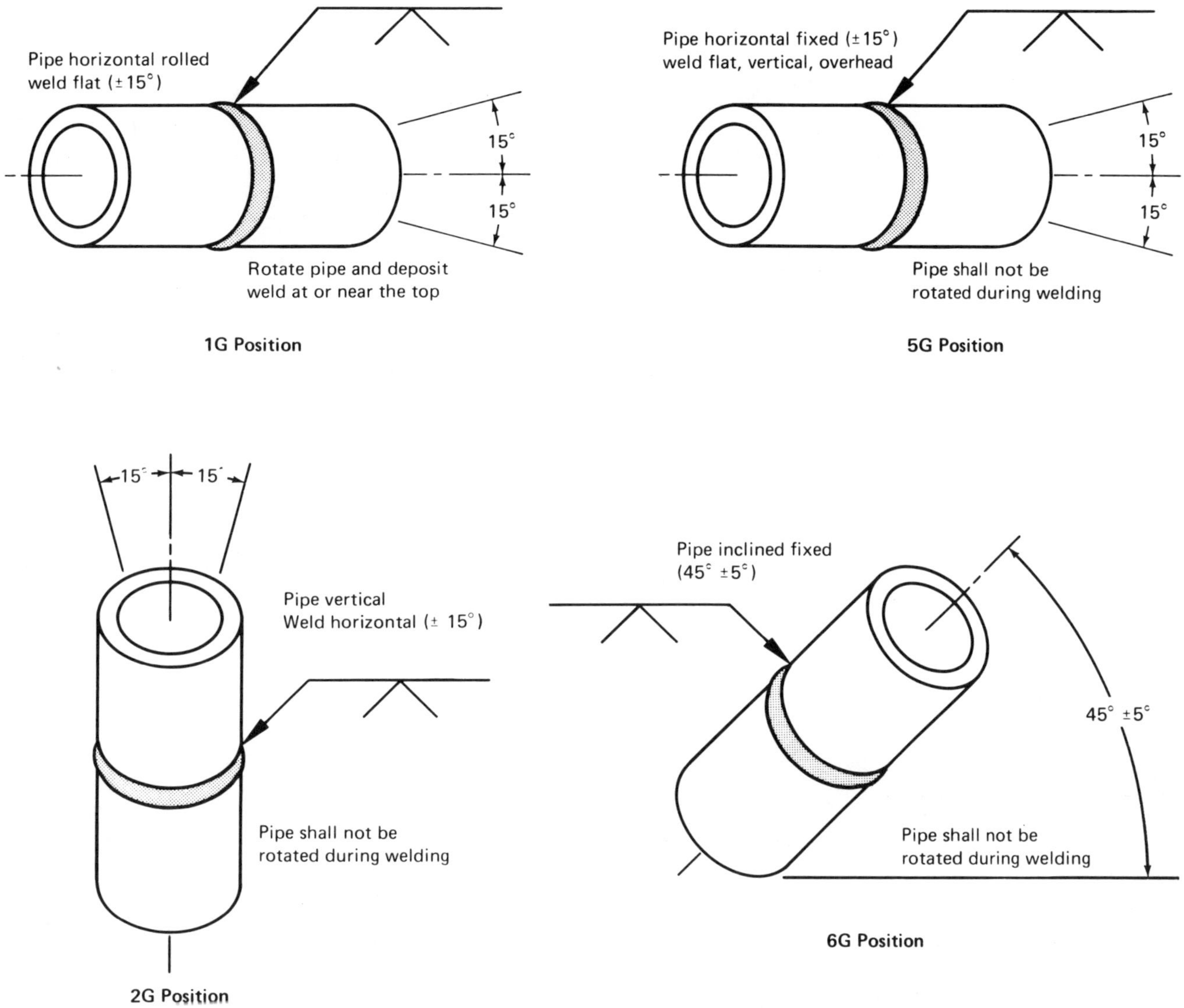


**Fig. 1.4.2–Test positions for groove welds in pipe**

for the welding of any fillet weld in pipe. If only fillet welds are required in production, qualification testing for fillet welds only may be conducted in accordance with Section 5. Qualification in one fillet weld position qualifies for welding in other positions as follows:

(1) Qualification in the 1F position qualifies only for this fillet weld position.

(2) Qualification in the 2F position qualifies for fillet welding in the 1F and 2F positions.

(3) Qualification in the 4F position qualifies for fillet welding in the 1F, 2F, and 4F positions only.

(4) Qualification in the 5F position qualifies for fillet welding in any position because it includes elements of all of those positions.

(5) Qualification in a position other than the four standard positions described above is valid only for that position (plus or minus 15 degrees).

**1.4.4** When production welding is to be performed in close quarters, the welding procedure and the welders should be qualified under similarly restrictive conditions.

**1.5 Definitions.** The nomenclature herein conforms to AWS A3.0, Terms and Definitions for Welding. Other terms not included in that document but commonly used in welding pipe and tube are defined in Appendix B. For the purpose of this document, the term 'contractor' refers to any organization undertaking to qualify welding procedures and welders in accordance with this specification.

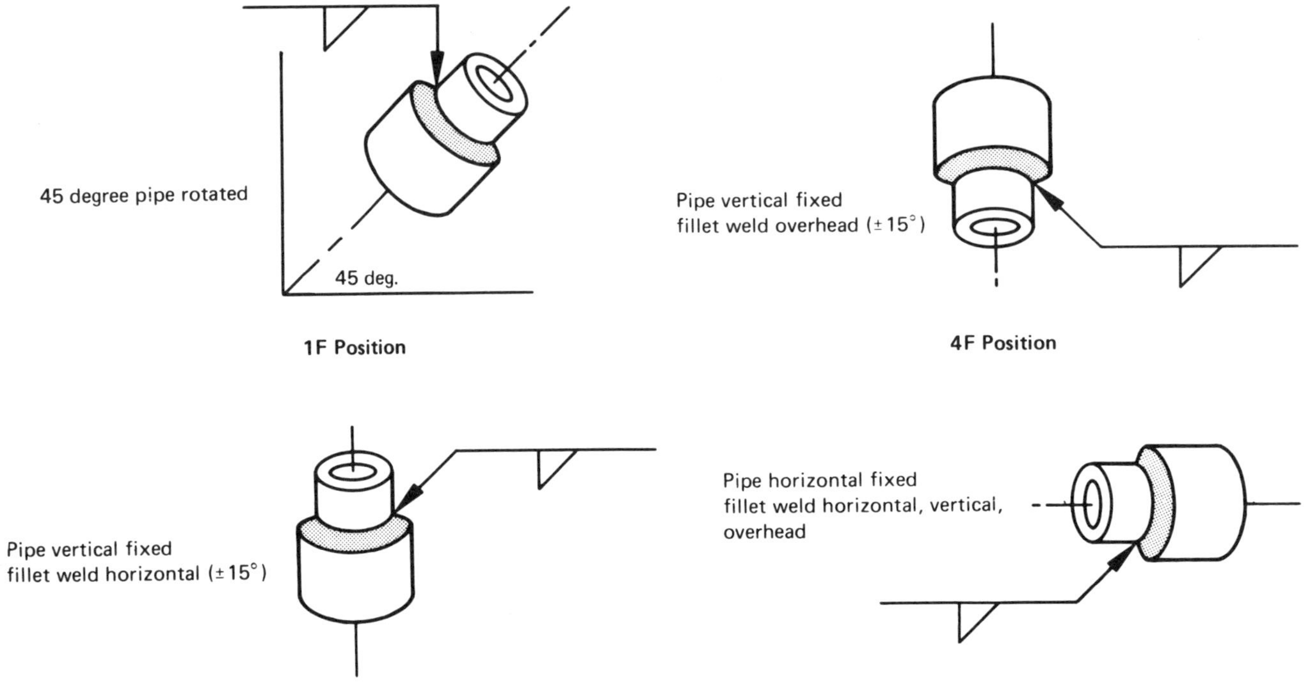


**Fig. 1.4.3—Test positions for fillet welds in pipe**

**1.6 Pipe Sizes.** For the convenience of the users of this specification, a table of data on commercially available pipe sizes, wall thicknesses, and schedules is included as Appendix E.

**1.7 Safety and Health.** The contractor shall ensure the safety and health of the welders and protect the workplace from fire and explosion by complying with the safety procedures recommended in ANSI Z49.1, Safety in Welding and Cutting, and in ANSI Z49.2/NFPA 51B, Safe Usage of the Cutting and Welding Processes.

## 2. AR-1 Level of Welding Procedure Qualification

**2.1 General Notes on Welding Procedure Qualification for Groove Welds**

Refer to Table 1 for sample welds and tests required for AR-1 welding procedure qualification. Prepare only those welds required to qualify the pipe or tube diameter-wall thickness ranges and positions of welding that will be involved in production welding.

If it becomes apparent that the welding conditions set forth in the tentative welding procedure will not produce a joint that meets the requirements, the test should be discontinued. One such indication is the need for an excessive number of repairs to produce an acceptable weld. In such a case, new conditions must be established and their validity proven by a new test weld.

Specimen preparation, testing techniques, and acceptance requirements for the tests prescribed in Table 1 are presented in the paragraphs following the table. The length of the sample welds should be sufficient to assure enough material to grip in the tensile test machine.

Paragraph 2.3 should be referred to prior to starting the procedure qualification testing to determine all of the variables that need to be controlled and documented. Also refer to the limits on changes allowable in welding variables controlling the qualification of an AR-1 welding procedure in paragraph 2.3 to determine when modification of an existing welding procedure requires the performance of new qualification tests.

**2.2 Tests Required for Sample Groove Welds.** Details regarding specimen preparation, testing techniques, and acceptance requirements for qualification of a welding procedure to the AR-1 level follow:

**2.2.1 Visual Inspection.** Each sample weld shall be inspected visually with the unaided eye:

(1) *Prior to assembly* of the test pieces, to assure that the proper materials are being used, that parts have been cleaned, and that the end preparation is correct.

(2) *After fit-up,* to check the joint clearances.

(3) *During welding,* to inspect for cracks, porosity, undercut, incomplete fusion, surface roughness, and poor crater conditions. Each weld bead should be inspected as deposited and after any normal mechanical cleaning prior to depositing subsequent passes. Brushing, filing, or grinding may be used when necessary on starts and stops,

## Table 1
### Sample welds and tests required for AR-1 level of welding procedure qualification

| Sample welds | | | | | Tests required on sample welds[a] | | | | | | |
|---|---|---|---|---|---|---|---|---|---|---|---|
| Pipe size of sample welds[b] | Pipe or tube[c] wall thicknesses qualified min, in. | Pipe or tube[c] wall thicknesses qualified max, in. | Number of sample welds per position | Pipe or tube size qualified, NPS, in.[a] | Visual inspection[d] | Penetrant inspection optional[e] | Radiography[f] optional | Tensile test[g], no. of specimens | Bend test[h], number of specimens Face | Root | Side |
| | | | | Wall thicknesses 0.750 in. and under | | | | | | | |
| 1/2 in. Sch. 40 | 0.065 | 0.400 | 1 | Through 1-1/2 | Yes | Yes | Yes | 1 | 0 | 0 | 0 |
| 2 in. Sch. 80 or 3 in. Sch. 40 | 0.065 | 0.674 | 1 | 1 through 4 | Yes | Yes | Yes | 1 | 1 | 1 | 0 |
| 6 in. Sch. 40 | 0.109 | 0.750 | 1 | Over 4 | Yes | Yes | Yes | 2 | 2 | 2 | 0 |
| | | | | Wall thicknesses over 0.750 in. | | | | | | | |
| 6 in., 8 in. or 10 in. Sch. 120 | 0.375 | Any | 1 | 6 and over | Yes | Yes | Yes | 2 | 0 | 0 | 4 |

a. See Appendix C for optional impact tests for low temperature service applications and microstructural examinations.
b. See Appendix C for option of job size pipe.
c. For pipe and tube having a wall thickness less than 0.065 in., individual qualification is required for each thickness by visual, penetrant, x-ray, and tensile testing.
d. See 2.2.1.
e. See 2.2.2. Required only if production welds are to be liquid penetrant or magnetic particle inspected.
f. See 2.2.3.
g. See 2.2.4 and 2.2.5.
h. See 2.2.4 and 2.2.6. Side bends shall be used instead of face and root bends for sample weld wall thicknesses of more than 0.375 in.

SI Equivalents

| in. | mm | in. | mm | in. | mm | in. | mm |
|---|---|---|---|---|---|---|---|
| 0.065 | 1.65 | 0.400 | 10.16 | 1/2 | 13 | 4 | 102 |
| 0.109 | 2.77 | 0.674 | 17.12 | 1-1/2 | 38 | 6 | 152 |
| 0.375 | 9.52 | 0.750 | 19.05 | 2 | 51 | 8 | 203 |
| | | | | 3 | 76 | 10 | 252 |

and to remove local defects appearing on the outer surface of the weld reinforcement. If production welding is to be accomplished from the outside of the pipe only, no repair of the root of the sample weld from the inside of the pipe is permitted.

(4) *After welding has been completed* (including any postheating), to detect any surface defects.

The following is required for a sample weld to pass Acceptance Requirements for visual inspection:

(a) The specified base metal and filler metal shall have been used.

(b) The weld shall be free of cracks.

(c) All craters shall be filled to the full cross section of the weld and shall be free of cracks upon completion of the welding.

(d) The face of the weld shall be at least flush with the outside surface of the pipe, and the weld shall merge smoothly with the base metal. Depth of weld undercut shall not exceed 1/32 in. (0.8 mm), irrespective of length. Weld reinforcement shall not exceed the following:

| Sample weld pipe size, in. | mm | Face reinforcement, in. | mm |
|---|---|---|---|
| 1/2 | 13 | 3/32 | 2.4 |
| 2 or 3 | 51-76 | 1/8 | 3.2 |
| 4 and over | 102 | 3/16 | 4.8 |

(e) The root of the weld shall be examined, and there shall be complete joint penetration and no cracks, nor incomplete fusion. A concave root surface is permitted within the limits shown in (f) below, provided the total weld thickness is equal to or greater than that of the base metal.

(f) A concave root surface and root reinforcement are permitted within the limits shown below:

**Limits of surface imperfections on the inside surface of the joint**

| Sample weld pipe size, in. | mm | Maximum root surface concavity, in. | mm | Maximum root reinforcement, in. | mm |
|---|---|---|---|---|---|
| 1/2 | 13 | 1/32 | 0.08 | 1/16 | 1.6 |
| 2 or 3 | 51-76 | 1/16 | 1.6 | 1/16 | 1.6 |
| 4 and over ⩽0.75 in. wall | 102 | 1/16 | 1.6 | 3/32 | 2.4 |
| 6 and over >0.75 in. wall | 152 | 1/16 | 1.6 | 1/8 | 3.2 |

**2.2.2 Liquid Penetrant or Magnetic Particle Inspection.** This test is required only for those welds that are to be subjected to this type of examination in production. Each sample weld shall be examined by liquid penetrant in accordance with Procedure A-2 or B-3 of ASTM E165. The weld shall be examined as required by the welding procedure, such as after the completion of the root pass or the entire weld, including any postheating. Dry powder magnetic particle examination in accordance with ASTM E109 may be used for magnetic materials in lieu of the liquid penetrant examination.

Acceptance requirements are as follows: Cracks are not acceptable; random porosity and slag not exceeding 1/32 in. (0.8 mm) in size are acceptable. No more than four indications 1/64 in. (0.4 mm) in size, or larger, in a line are acceptable when they are less than 1/16 in. (1.6 mm) apart.

**2.2.3 Radiographic Examination.** The weld reinforcement on the outside surface of each sample weld shall be mechanically removed to the extent that surface ripples cannot mask or be confused with the image of any internal defects in the radiographs. The weld reinforcement may be removed flush with the surface of the pipe, but if it is only partially removed, its height shall not exceed 3/32 in. (2.4 mm). No removal of metal on the inside surface of the pipe is permitted, except for sample welds representing large diameter pipe where the inside surface of production welds will also be ground or back-welded.

**2.2.3.1** General radiographic techniques are as follows:

(1) All completed sample welds shall be radiographed for their full circumference.

(2) In order to obtain the required sensitivity, it is usually necessary to use an x-ray machine as the source of radiation for sample welds in pipe 3-1/2 in. (89 mm) and under. Either an x-ray machine or a radioactive source may be used for sample welds in pipe 4 in. (102 mm) and over, except for aluminum pipe, where only an x-ray machine should be used for all thicknesses.

(3) A penetrameter shall be used to verify the adequacy of radiographic technique. The penetrameter shall conform to the requirements of ASTM E142, level 2-2T, or any other equivalent penetrameter giving 2 percent sensitivity.

(4) The material of the penetrameter shall be radiographically similar to the filler metal. The penetrameter is normally placed over the base metal near the weld, but if the filler metal is not radiographically similar to the base metal, then the penetrameter may be placed over the weld metal.

(5) The weld shall be radiographed with a technique having sufficient sensitivity to reveal internal defects and indicate the features of the penetrameter.

(6) Each penetrameter shall represent an area of essentially uniform radiographic density as judged by density comparison strips or a densitometer. An additional penetrameter shall be required if the film density varies more than −15 or +30 percent from the density through the penetrameter. The film density through acceptable weld metal shall be 2.0 minimum for single film viewing.

(7) The images of the identifying numbers, the penetrameter outline, and the 2T hole are all essential indexes of image quality, and they shall appear clearly on the radiograph. The difference in density between the images of the holes and the penetrameter shall be the

same as that between the edge of the penetrameter and the adjacent film area.

(8) If the reinforcement of the backing is not removed, a shim of a material radiographically similar to the filler metal shall be placed under the penetrameter. The thickness of the shim shall be the same as the reinforcement and any backing not removed.

(9) Identification markers, the images of which shall appear on the film, shall be placed on the base metal adjacent to the weld, not on the film, and their locations shall be marked on the surface near the weld in such a manner that a defect appearing on a radiograph may be accurately located. The markers shall be located in a manner that will indicate on the film that complete radiographic coverage of the weld has been obtained.

(10) The following shall appear on each radiograph or in the records accompanying each radiograph: identity of the organization making the sample weld, the date, test weld number, welder's stamp number, welding procedure specification number, and material designation.

(11) All radiographs shall be free from excessive mechanical, chemical, or other processing defects that could interfere with proper interpretation of the radiograph.

2.2.3.2 Specific radiographic techniques for sample welds are as follows:

(1) *Small (1/2 to 3-1/2 in. [13 to 89 mm]) diameter sample welds.* A technique shall be used in which the radiation passes through two walls of the pipe at such an angle that the weld in each wall is viewed for acceptance on the same film. Three views are required approximately 120 degrees apart.

The penetrameter shall be on the source side of the outside surface of the pipe with the long dimension parallel to the axis of the pipe, and it shall be positioned so that it does not interfere with the film interpretation of any portion of the weld. The penetrameter thickness shall be based on twice the pipe wall thickness.

(2) *4, 5, and 6 in. (102, 127, and 152 mm) sample welds.* A technique shall be used in which the radiation passes through two walls of the pipe at an angle, but only the portion of the weld closest to the film shall be viewed for acceptance. Four views are required.

The penetrameter shall be on the source side of the outside surface of the pipe with the long dimension parallel to the axis of the pipe, and it shall be positioned so that it does not interfere with the film interpretation of any portion of the weld. The penetrameter thickness shall be based on twice the pipe wall thickness.

(3) *8 in. (203 mm) and larger sample welds.* The weld shall be radiographed using a technique in which the source is either inside or outside the pipe, and the radiation passes through only one wall to reach the film.

When the source is outside the pipe, six views are required. The penetrameter shall be placed on the source side in all cases and shall be normal to the radiation. The long dimension of the penetrameter shall be parallel to the long axis of the weld. The penetrameter thickness shall be based on one pipe wall thickness.

2.2.3.3 Sample welds having any of the following defects shall be judged unacceptable:

(1) Any type of crack

(2) Incomplete fusion between the weld metal and base metal

(3) Incomplete fusion between any of the weld passes

(4) Lack of complete joint penetration

(5) Slag inclusions greater than those listed in the table below

**Limits of permissible slag inclusions**

| Sample weld pipe size, in. | mm | Length of slag inclusion in the full circumference of the weld, max, in. | mm |
|---|---|---|---|
| 1/2 | 12.7 | 1/8 Single | 3.2 |
| 2 or 3 | 51-76 | 3/16 Single | 4.8 |
| | | 1/2 Total | 13.0 |
| 6-8 | 152-203 | 1/4 Single | 6.4 |
| | | 1/2 Total | 13.0 |
| 10 and over | 254 | 1/4 Single | 6.4 |
| | | 3/4 Total | 19.0 |

(6) Porosity and tungsten inclusions in excess of the amounts shown in Fig. 2.2.3.3(6)

**2.2.4 Preparation of Sample Welds for Destructive Tests**

**2.2.4.1 1/2 in. (13 mm) Sample Welds.** Prepare the sample weld for the full size tensile test by removing the weld reinforcement until it is flush with the base metal.

**2.2.4.2 2 or 3 in. (51 or 76 mm) Sample Welds.** Remove and prepare from the sample weld one tensile, one face-bend, and one root-bend specimen as shown.

**2.2.4.3 6, 8, and 10 in. (152, 203, and 254 mm) Sample Welds.** Remove and prepare from the sample weld two tensile, two face-bend, and two root-bend specimens as shown. Four side-bend specimens may be substituted for the face-bend and root-bend specimens for wall thicknesses of 3/8 in. (9.5 mm).

**2.2.5 Tensile Tests.** The tensile specimen shall be ruptured under tensile load. The tensile strength shall be computed by dividing the maximum load by the least cross-sectional area of the specimen as measured before the load is applied. The 1/2, 3/4, 1, and 1-1/2 in. (13, 19, 25, and 38 mm) full pipe size sample welds shall be tested using the fixture shown in Fig. 2.25 after the outside diameter and the wall thickness have been measured.

In order to pass the tensile test, the specimen shall have a tensile strength not less than the minimum tensile strength value given in column 2 of Table 2 for the material involved. The tensile strength values specified are for welded assemblies. For welds of dissimilar metals, the specimen shall have a tensile strength not less than the value given for the weaker of the two materials.

Permissible porosity indications for various weld thickness are illustrated in the following charts. The test radiographs shall be compared with these charts visually to form a basis for acceptance or rejection. In cases of dispute, the detailed requirements in the following paragraphs will govern:

(1) Dark images of a generally circular or oval shape shall be interpreted as porosity. Light images of a generally circular nature shall be interpreted as tungsten inclusions in gas tungsten arc welds. For simplicity, porosity and tungsten inclusions are called "pores" in the charts, and the combined indications shall not exceed the number and size shown in the charts.

(2) The maximum pore dimensions shall be 20 percent of t, where t is the wall thickness of the sample weld, or 1/8 in. (3.2 mm), whichever is smaller.

(3) The following porosity charts illustrate various types of assorted and uniform, randomly dispersed porosity indications. These charts for each thickness represent the maximum acceptable porosity. The charts represent full-scale six inch (152 mm) radiographs and shall not be enlarged or reduced. The porosity distributions shown are not necessarily the patterns that may appear on the radiographs but are typical of the number and size of indications permitted.

(4) Aligned porosity shall be acceptable, providing the summation of the diameters of the pores is no more than T in a length 12T, or 1/2 in. (12.7 mm) in 6 in. (152 mm), whichever is less, providing each pore is separated by a distance of at least six times the diameter of the largest adjacent pore.

(5) Permissible porosity indications for weld thicknesses intermediate to those illustrated may be evaluated by comparison with the next thinner material.

**1/2 in., 1 in., and 1-1/2 in. Sch. 40 sample welds**

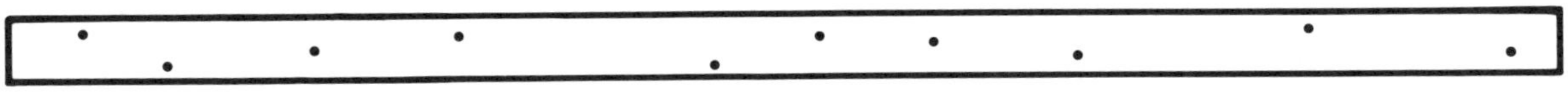

**2 in. Sch. 80 or 3 in. Sch. 40 sample welds**

Assorted

Large

Medium

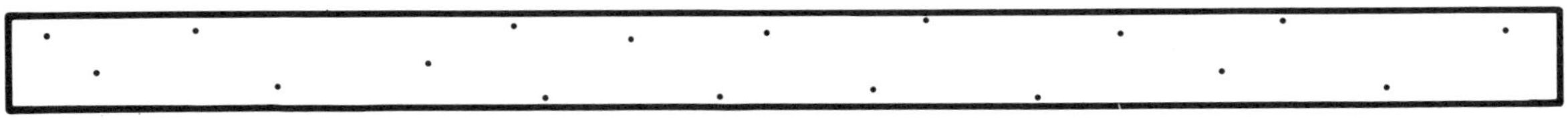

Fine

**6 in. Sch. 40 sample welds**

**Fig. 2.2.3.3 (6)—Porosity charts for radiographically-determined porosity and tungsten inclusions in welds**

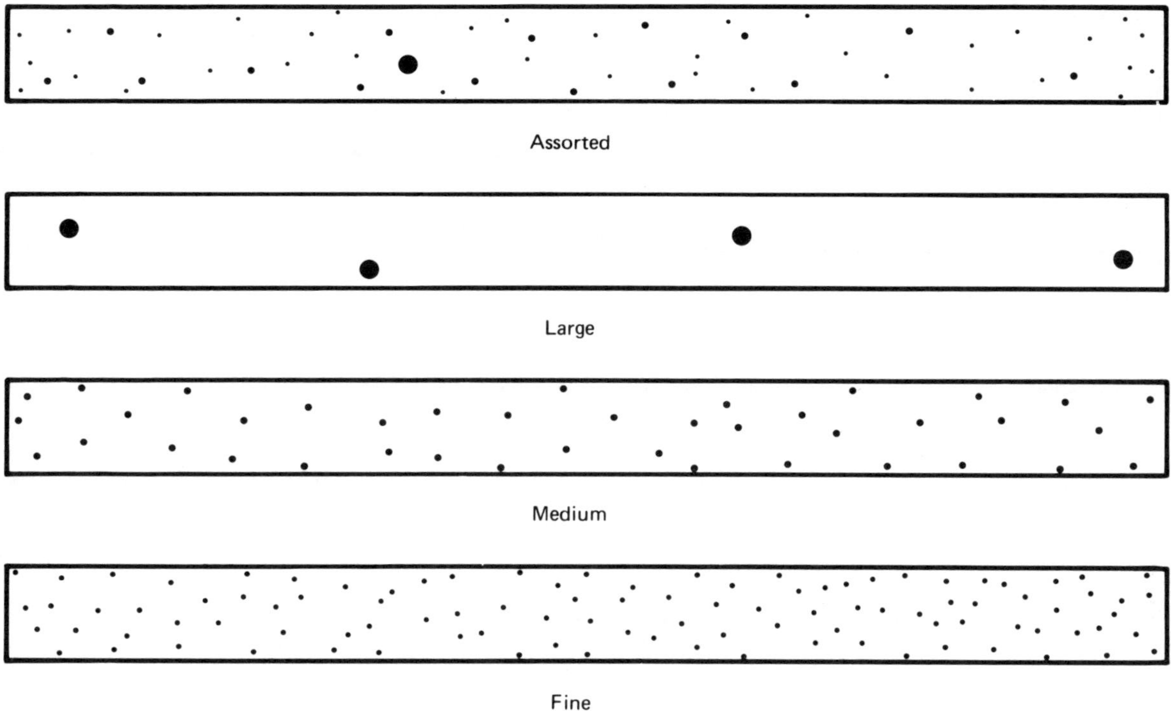


8 in. Sch. 80 and 10 in. Sch. 40 sample welds

**Fig. 2.2.3.3 (6) (continued)–Porosity charts for radiographically-determined porosity and tungsten inclusions in welds**

**2.2.6 Guided-Bend Test**

(1) Bend specimens shall be tested in a guided-bend jig or the wraparound jig of the general type shown in Fig. 2.2.6.

(2) The B and D dimensions of the guided-bend jig and the B dimension of the wraparound jig shall be as follows, depending upon the material and thickness:

**Sample thicknesses and jig dimensions[a,b]**

| (See Table 2) | Sample thickness, in. | mm | Specimen thickness, in. | mm | Radius of bend, B, in. | mm | Radius of die, D, in. | mm |
|---|---|---|---|---|---|---|---|---|
| All except P23 | t<br>0.375 | <br>9.5 | t<br>0.375 | <br>9.5 | 3-1/3t<br>1-1/4 | <br>32.0 | 4-1/3t − 1/16<br>1-11/16 | 1.6<br>43.0 |
| P23 (and F23) | t < 0.125<br>t > 0.125 | 3.2<br>3.2 | t<br>0.125 (face-, root-, or side-bend) | <br>3.2 | 8-1/4t<br>1-1/32 | <br>26.0 | 9-1/4t + 1/16<br>1-7/32 | 1.6<br>31.0 |

a. t = the wall thickness of the sample weld.
b. For sample welds in P23 materials whose thickness exceeds 0.125 in. (3.2 mm), remove and bend specimens of 0.125 in. (3.2 mm) thickness as required to cover the wall thickness under test.

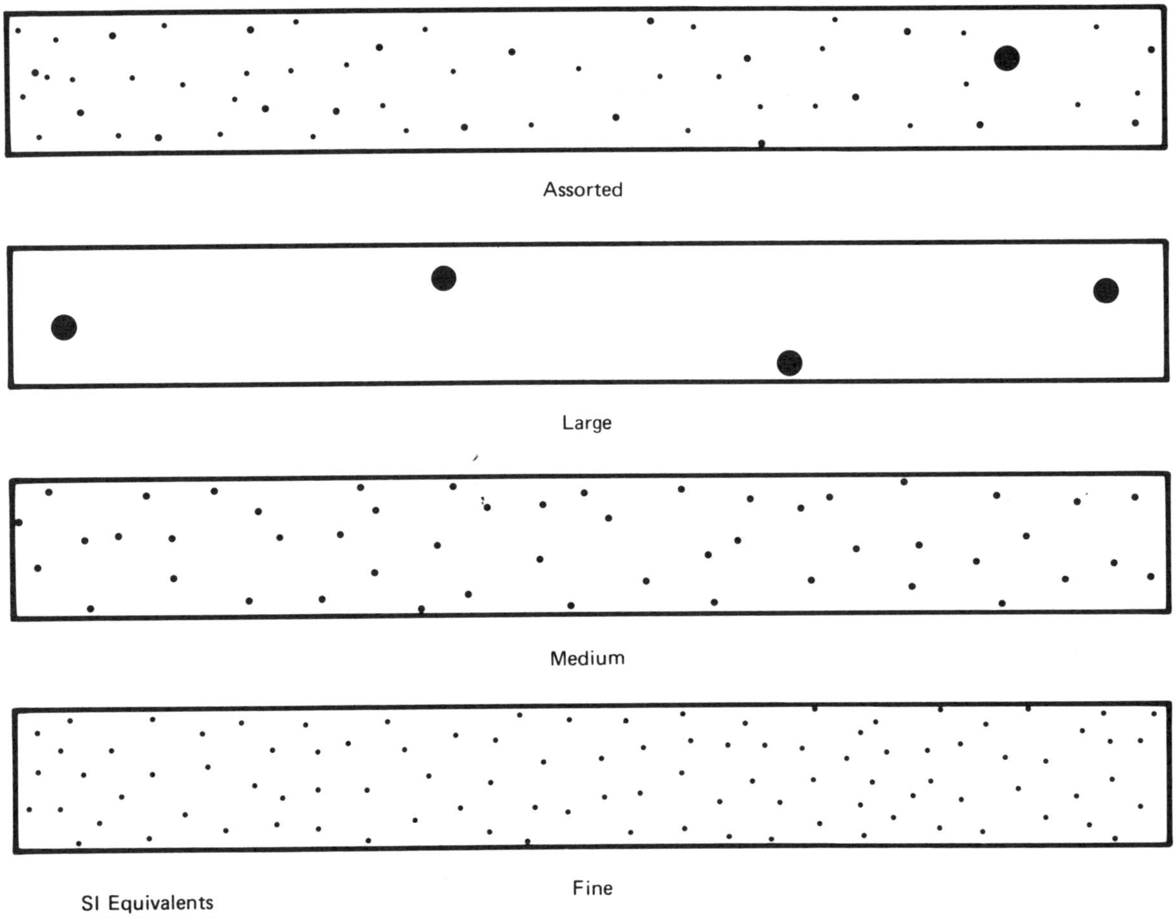


8 in. and 10 in. Sch. 120 sample welds

SI Equivalents

| in. | mm |
|---|---|
| 1/2 | 13 |
| 1 | 25 |
| 1-1/2 | 38 |
| 2 | 51 |
| 3 | 76 |
| 6 | 152 |
| 8 | 203 |
| 10 | 254 |

**Fig. 2.2.3.3 (6) (continued)—Porosity charts for radiographically-determined porosity and tungsten inclusions in welds**

(3) When using the guided-bend jig, the specimen shall be placed on the die with the weld at midspan. Face-bend specimens shall be placed with the face of the weld directed toward the gap; root-bend specimens shall be placed with the root of the weld directed toward the gap, and side-bend specimens with the surface having the greater defects, if any, toward the gap.

(4) The plunger shall force the specimen into the die until the specimen becomes U-shaped and a 1/8 in. (3.2 mm) diameter wire cannot be inserted between the die and the specimen. The weld and heat-affected zones shall be centered and completely within the bent portion of the specimen after testing.

(5) When using the wraparound jig, the specimen shall be firmly clamped on one end so that there is no sliding of the specimen during the bending operation. The weld and heat-affected zones shall be completely in the bent portion of the specimen after testing. Test specimens shall be removed from the jig when the outer roll has been moved 180 degrees from the starting point.

In order to pass the acceptance requirements for the test, the bent specimens shall have no cracks or other open defects exceeding 1/8 in. (3.2 mm) measured in any direction on the convex surface of the specimen, except that cracks occurring at the corners shall not be considered unless there is definite evidence that they result from slag inclusions or other internal defects.

*Note:* Details for optional bend tests are described in Appendix C.

**2.3 Limitations on Changes Allowable in Welding Variables Controlling the Qualification of a Welding Procedure to the AR-1 Level**

Information concerning all of the following applicable welding variables shall be recorded when a welding procedure is first qualified. Changes, other than those given below, may be made without the necessity for qualifying a new welding procedure, provided the existing welding procedure specification is revised to show these changes. The welding procedure shall be requalified and set up as a new procedure when any of the changes listed below are made in the welding variables.

**2.3.1 Base Metal.** The grouping of materials by P-number classification is based on chemical and metallurgical characteristics. The grouping of filler metals by F-number classification is based on usability characteristics. These groupings are intended to reduce the number of welding procedure qualifications required.

(1) A change from a base metal listed under one P-number in Table 2 to material listed under another P-number, or to any other base metal, shall require requalification.

(2) A change from a base metal listed under Group 9A to one listed under Group 9B in P-number 9 shall require requalification (see footnotes in Table 2).

(3) A change from a base metal listed under one letter subgroup to one listed under another subgroup in P-number 10 shall require requalification.

(4) A change from a similar metal joint to a dissimilar metal joint (two different P-numbers), even though welding procedures have been developed for each of the metals welded to itself, shall require requalification.

**2.3.2 Pipe Size and Wall Thickness.** A change from one diameter wall grouping to another, as shown in Table 1, shall require requalification.

**2.3.3 Joint Design.** A change from a single-V-groove or a single-U-groove to a single-bevel-groove, single-J-groove, or square-groove; or from a single-bevel-groove or a single-J-groove to a square-groove shall require requalification.

An increase or decrease of 1/16 in. (1.6 mm) or more in root opening and root-face dimensions beyond those specified in the welding procedure shall require requalification.

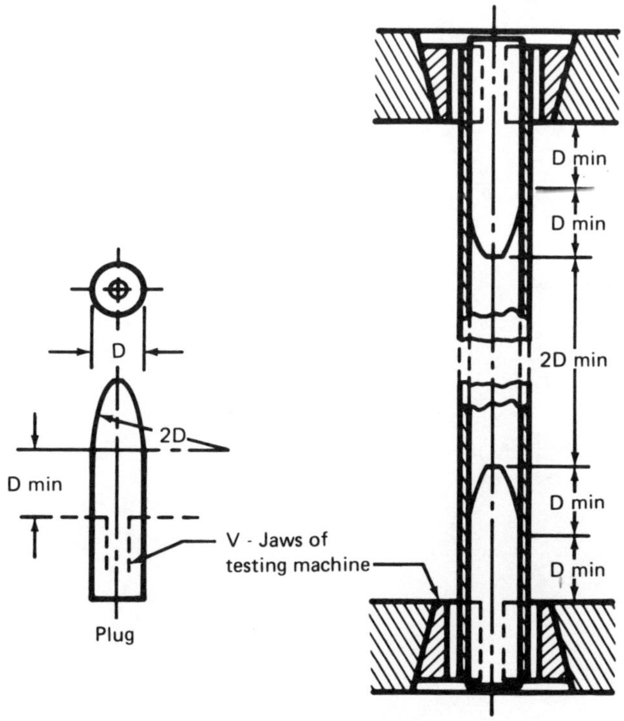


**Tensile testing fixture**

**Fig. 2.2.5—Details of destructive tests**

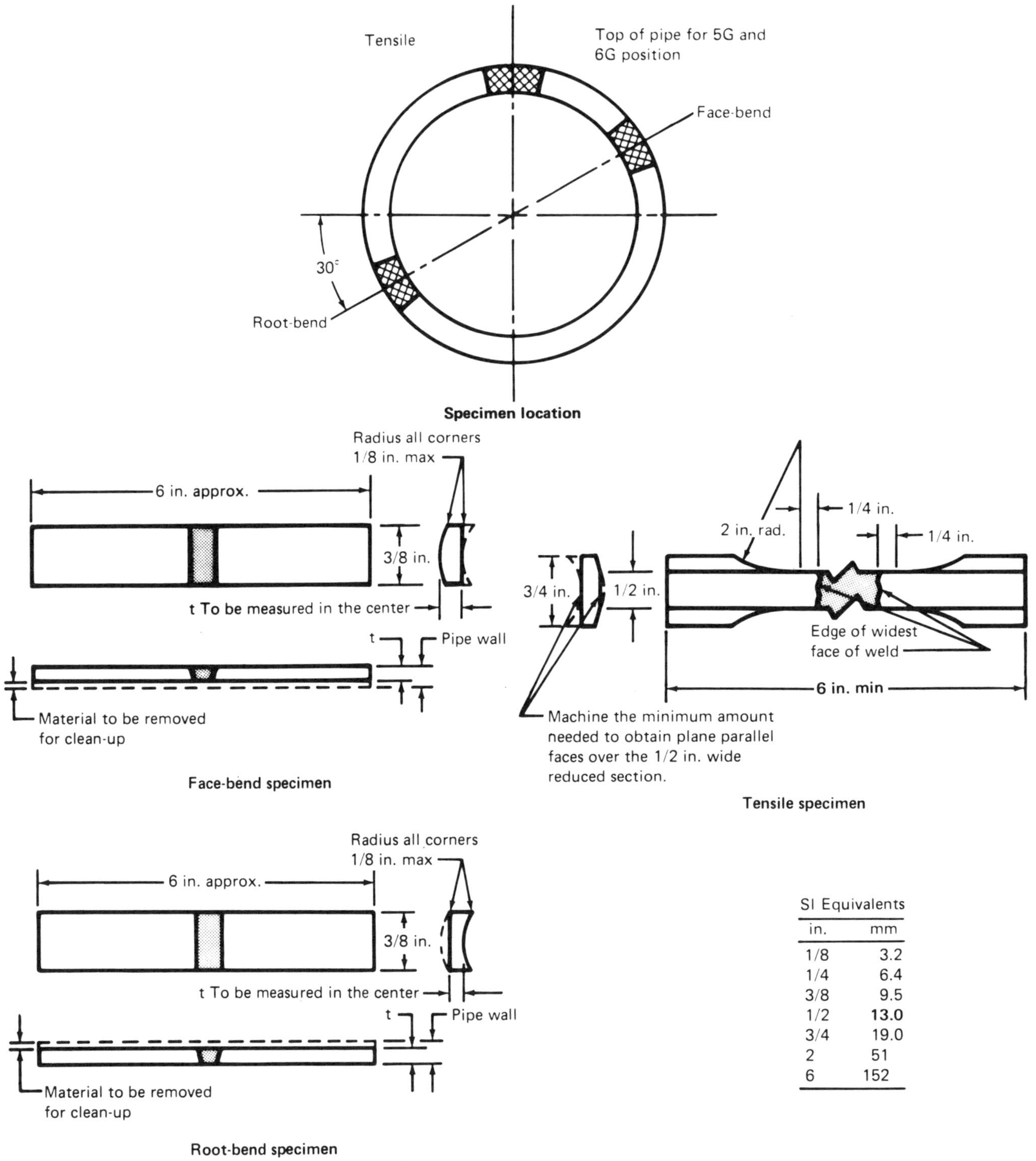


SI Equivalents

| in. | mm |
|---|---|
| 1/8 | 3.2 |
| 1/4 | 6.4 |
| 3/8 | 9.5 |
| 1/2 | **13.0** |
| 3/4 | 19.0 |
| 2 | 51 |
| 6 | 152 |

Notes:

1. All dimensions in inches.
2. Weld reinforcement and backing, if any, shall be removed flush with the surface of the specimen. Cut surfaces shall be smooth and parallel.

**Fig. 2.2.5 (continued)–Details of destructive tests**

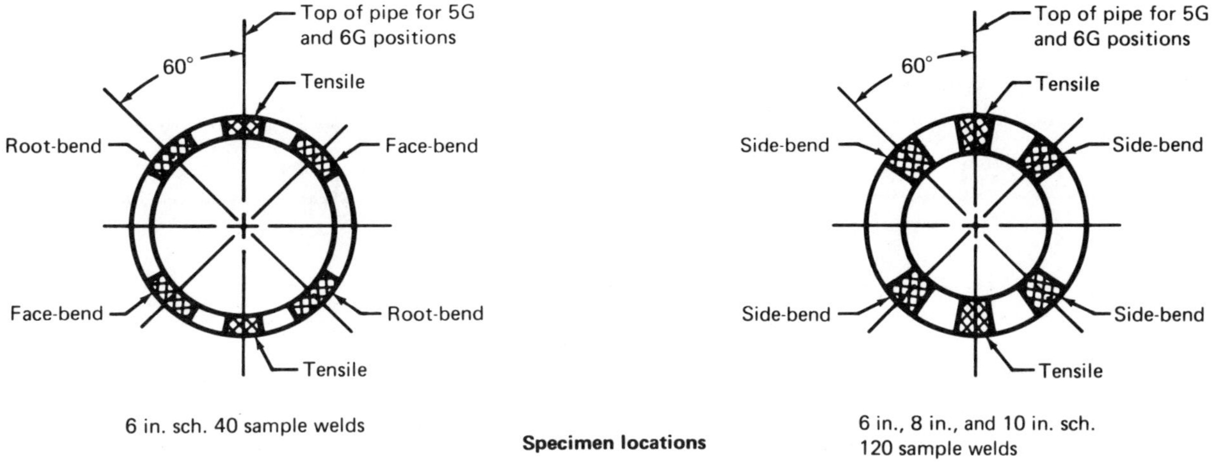


**Specimen locations**

Pipe wall 3/8 in. or less

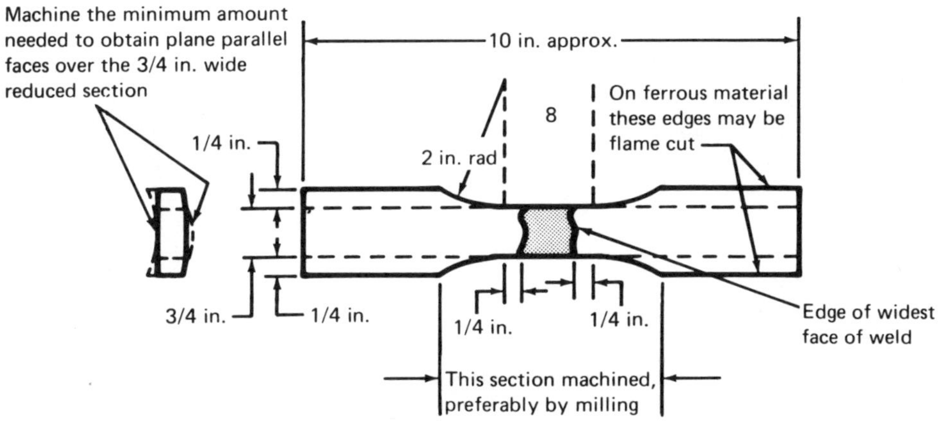


**Tensile specimen**

| SI Equivalents | |
|---|---|
| in. | mm |
| 1/4 | 6.4 |
| 3/4 | 19.0 |
| 2 | 51 |
| 6 | 152 |
| 8 | 203 |
| 10 | 254 |

Notes:

1. All dimensions in inches.
2. Weld reinforcement and backing, if any, shall be removed flush with the surface of the specimen. Cut surfaces shall be smooth and parallel.

**Fig. 2.2.5 (continued)—Details of destructive tests**

**2.3.4 Backing** (for complete joint penetration groove welds only).

(1) The omission of a backing ring for welds made from one side only shall require requalification.

(2) For copper alloys, the addition of a backing ring shall require requalification.

(3) The addition or omission of a preplaced consumable insert shall require requalification.

(4) The omission of a removable material as a backing device shall require requalification.

(5) The addition or omission of a gas or flux as a means of backing shall require requalification.

(6) The omission of a backing weld shall require requalification.

**2.3.5 Filler Metal**

**2.3.5.1 Composition.** For ferrous materials, a change in the chemical composition of the weld deposit from one A-number to any other A-number shown in Table 3 requires requalification.

Filler metal or deposited weld metal compositions other than those listed in Tables 3, 4, and 5 shall require separate qualification.

**2.3.5.2 Usability.** For both ferrous and nonferrous filler metals, a change from one F-number to another F-number in Tables 4 and 5 requires requalification.

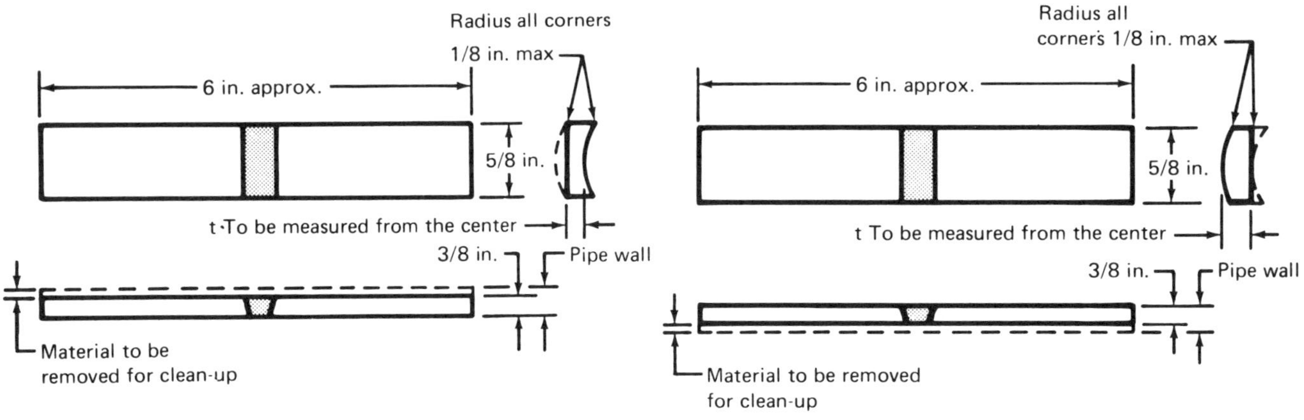


**Root-bend specimen** **Face-bend specimen**

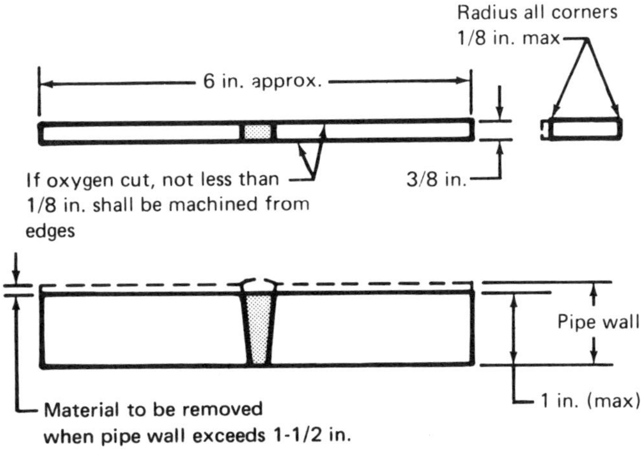


**Side-bend specimen**
(for pipe over 3/8 in. thick)

SI Equivalents

| in. | mm |
|---|---|
| 1/8 | 3.2 |
| 3/8 | 9.5 |
| 5/8 | 15.9 |
| 1 | 25.4 |
| 6 | 152 |

Notes:

1. All dimensions in inches.
2. Weld reinforcement and backing, if any, shall be removed flush with the surface of the specimen. Cut surfaces shall be smooth and parallel.

**Fig. 2.2.5 (continued)—Details of destructive tests**

**2.3.6 Welding Process.** A change from one welding process to another, for making either the entire weld or part thereof, shall require requalification. Within a given process, a change from fully automatic to semiautomatic or manual or a change from semiautomatic to manual (or vice versa) shall require requalification. Additional requirements or exceptions for various welding processes are listed below.

**2.3.6.1 Oxyacetylene or Oxyhydrogen Process**

(1) The maximum thickness qualified using the oxyacetylene or oxyhydrogen welding process for pipe 4 in. (102 mm) and under is the thickness of the sample weld plus 20 percent.

(2) For pipe sizes over 4 in. (102 mm), the maximum pipe size and wall thickness qualified are the sample weld pipe size and wall thickness plus 20 percent.

(3) The addition of backing rings or backing material shall require requalification.

**2.3.6.2 Gas Tungsten Arc Process**

(1) The addition or omission of filler metal shall require requalification.

(2) The omission of high frequency stabilization shall require requalification.

(3) For single-pass welding, the maximum thickness welded in production shall not exceed the thickness of the test pipe.

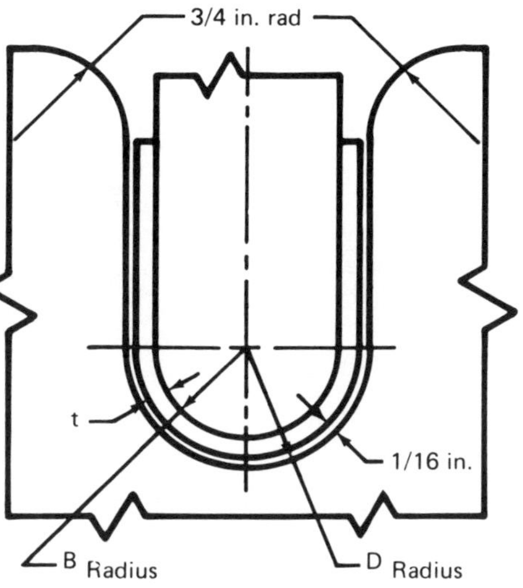


Schematic view of guided-bend jig

| SI Equivalents | |
|---|---|
| in. | mm |
| 1/16 | 3.2 |
| 3/4 | 19.0 |

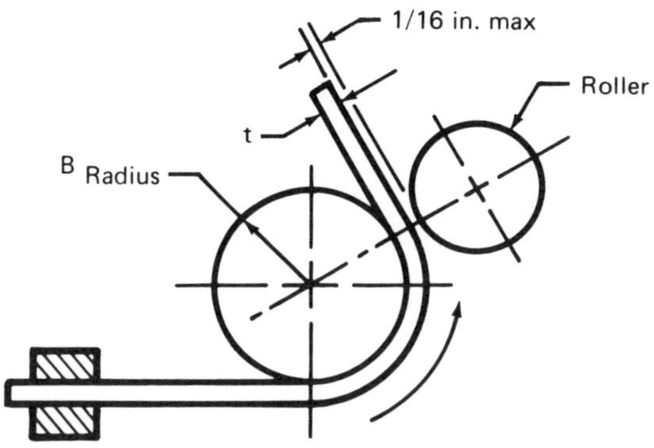


Schematic view of wraparound jig

Note: All dimensions in inches.

**Fig. 2.2.6—Jigs for guided-bend tests**

**2.3.6.3 Gas Metal Arc Process**

(1) For single-pass welding, the maximum thickness welded in production shall not exceed the thickness of the test pipe.

(2) A change from multipass to single-pass welding shall require requalification except for single-pass fillet welds not exceeding 1/4 in. (6.4 mm), which do not require requalification.

(3) A change in the filler metal from bare solid electrodes to metal cored electrodes, or vice versa, shall require requalification.

**2.3.6.4 Flux Cored Arc Process**

(1) For single-pass welding, the maximum thickness welded in production shall not exceed the thickness of the test pipe.

(2) A change from multipass to single-pass welding shall require requalification except for single-pass fillet welds not exceeding 1/4 in. (6.4 mm), which do not require requalification.

**2.3.6.5 Submerged Arc Process**

(1) A change in the nominal composition of flux shall require requalification.

(2) In Table 3, a change from one type of weld deposit to any other shall require requalification, except that qualification with A-1 shall qualify the procedure for welding with A-2 and vice versa.

(3) Any change that would result in the important alloying elements in the weld metal being outside the composition range given in the welding procedure shall require requalification.

(4) A change from a single arc to multiple arc, or vice versa, or a change from one type of multiple arc to another type, such as from series arc to parallel arc, shall require requalification.

(5) A change from multiple-pass to single-pass welding shall require requalification. For single-pass welding, the maximum thickness welded in production shall not exceed the thickness of the test pipe.

**2.3.7 Gas Shielding.** In gas metal arc, gas tungsten arc, or flux cored arc welding, a change in the composition of the gas involving one or more of the following shall require requalification:

(1) A change from one gas to another.

(2) The addition or omission of an active gas such as oxygen, hydrogen, or carbon dioxide.

(3) The addition or omission of more than 5 percent by volume of another inert gas.

(4) With gas mixtures, a 5 percent change in the volume of one of the gases.

(5) A decrease of 10 percent or more in the gas flow rate.

**2.3.8 Electrical Characteristics**

(1) A change in the welding current from AC to DC, or vice versa, shall require requalification.

(2) A change in polarity shall require requalification.

(3) For the gas metal arc process, a change in current density that results in a change from spray arc, globular arc, or pulsating arc to a short circuiting arc, or vice versa, shall require requalification.

## Table 2
## Grouping of materials for qualification for the AR-1 level

The grouping of materials in this table of P-number classifications is based on the similarity of chemical and metallurgical characteristics to reduce the number of welding procedure qualifications.

Several P-number classifications have undergone additional subdivisions by the use of *group numbers.* For carbon steel (P-number 1), these group numbers are based on minimum tensile strength. Whenever group numbers have been assigned to other P-numbers, such assignments have been based on similarity of chemical analysis. This subdivision relates to the selection of filler materials and postweld heat treatment practices. To meet other metallurgical requirements, such as notch toughness, the properties of each specific base metal must be considered.

While most tubing and piping materials have been grouped below, only a limited number of other product forms have been included. Materials not listed below (including such items as piping, forgings, castings, and plates) may be grouped for qualification purposes with the piping and tubing materials having comparable technical, mechanical, and metallurgical characteristics.

| ASTM material specification | | Tensile strength, min, ksi | MPa | Type of material |
|---|---|---|---|---|
| P-number 1, Group 1[a] (Minimum tensile strength ⩽ 60 ksi [415 MPa]) | | | | |
| **Carbon steel** | | | | |
| A53 | Openhearth | 45 | 310 | Furnace welded pipe |
| | Grade A | 48 | 330 | Seamless or welded pipe |
| | Grade B | 60 | 415 | Seamless or welded pipe |
| A106 | Grade A | 48 | 330 | Seamless pipe |
| | Grade B | 60 | 415 | |
| A120 | | 45 | 310 | Galvanized pipe |
| A135 | Grade A | 48 | 330 | Electric resistance welded pipe |
| | Grade B | 60 | 415 | |
| A139 | Grade A | 48 | 330 | Fusion welded pipe |
| | Grade B | 60 | 415 | |
| | Grade C | 60 | 415 | |
| | Grade D | 60 | 415 | |
| A161 | Low carbon | 47 | 325 | Low carbon tubes |
| A178 | Grades A, C | 60 | 415 | Electric resistance welded tubes |
| A181 | Grade 1 | 60 | 415 | Forgings and fittings |
| A192 | | 47 | 325 | Seamless tubes |
| A210 | Grade A1 | 60 | 415 | Seamless tubes |
| A226 | | 47 | 325 | Electric resistance welded tubes |
| A333 | Grade 1 | 55 | 380 | Pipe for low temperature service |
| | Grade 6 | 60 | 415 | |
| A334 | Grade 1 | 55 | 380 | Tubes for low temperature service |
| | Grade 6 | 60 | 415 | |
| A381 | Class Y35 | 60 | 415 | Line pipe |
| | Class Y42 | 60 | 415 | |
| A500 | Grade A | 45 | 310 | Structural tubing |
| | Grade B | 58 | 400 | |
| A501 | | 58 | 400 | Structural tubing |

**Table 2**
**Grouping of materials for qualification for the AR-1 level (cont'd)**

| ASTM material specification | | Tensile strength, min, ksi | MPa | Type of material |
|---|---|---|---|---|
| P-number 1, Group 1[a] | | | | |
| **Carbon steel** | | | | |
| A523 | Grade A | 48 | 330 | Conduit pipe |
| | Grade B | 60 | 415 | |
| A524 | Grade I | 60 | 415 | Seamless pipe |
| | Grade II | 55 | 380 | |
| A539 | | 45 | 310 | Coiled tubing |
| A556 | Grade A2 | 47 | 325 | Seamless tubes |
| | Grade B2 | 60 | 415 | |
| A557 | Grade A | 47 | 325 | Electric resistance welded tubes |
| | Grade B | 60 | 415 | |
| A587 | | 48 | 330 | Low carbon pipe |
| A660 | Grade WCA | 60 | 415 | Centrifugally cast pipe |
| API 5L | Grade A | 48 | 330 | Line pipe |
| | Grade B | 60 | 415 | |
| API 5LS | Grade A | 48 | 330 | Spiral weld line pipe |
| | Grade B | 60 | 415 | |
| API 5LX | Grade X42 | 60 | 415 | Spiral weld line pipe |
| | Grade X42 | 60 | 415 | High-test line pipe |
| P-number 1, Group 2[a] | | | | |
| **Carbon steel** | | | | |
| A105 | | 70 | 480 | Forgings and fittings |
| A106 | Grade C | 70 | 480 | Seamless pipe |
| A139 | Grade E | 66 | 455 | Fusion welded pipe |
| A181 | Grade II | 70 | 480 | Forgings and fittings |
| A210 | Grade C | 70 | 480 | Seamless tubes |
| A381 | Class Y46 | 63 | 435 | Line pipe |
| | Class Y48 | 67 | 460 | |
| | Class Y50 | 69 | 475 | |
| A500 | Grade C | 62 | 430 | Structural tubing |
| A556 | Grade C2 | 69 | 480 | Seamless tubes |
| A557 | Grade C | 70 | 480 | Resistance welded tubes |
| A595 | Grade A | 65 | 450 | Low carbon tapered structural tubing |
| | Grade B | 70 | 480 | |
| A618 | Grade I | 70 | 480 | High-strength low alloy structural tubing |
| | Grade II | 70 | 480 | |
| | Grade III | 65 | 450 | |
| A660 | Grade WCB | 70 | 480 | Centrifugally cast pipe |
| | Grade WCC | 70 | 480 | |
| API 5LS | Grade X46 | 63 | 435 | Spiral weld line pipe |
| | Grade X52 | 66 | 455 | |
| API 5LX | Grade X42 | 60 | 415 | High-test line pipe |
| | Grade X46 | 63 | 435 | |
| | Grade X52 | 66 | 455 | |

**Table 2**
**Grouping of materials for qualification for the AR-1 level (cont'd)**

| ASTM material specification | | Group no. | Tensile strength, min, ksi | Tensile strength, min, MPa | Type of material |
|---|---|---|---|---|---|
| P-number 1, Group 3[a] | | | | | |
| **Carbon steel** | | | | | |
| A381 | Class Y52 | | 72 | 495 | Line pipe |
| API 5LX | Grade X60 | | 75 | 515 | Line pipe |
| | Grade X65 | | 80 | 550 | |
| | Grade X70 | | 82 | 565 | |
| API 5LS | Grade X56 | | 71 | 490 | Spiral weld line pipe |
| | Grade X60 | | 75 | 515 | |
| | Grade X65 | | 77 | 530 | |
| | Grade X70 | | 82 | 565 | |
| P-number 3 | | (Group 1 - Nominal composition: C-1/2 Mo) (Group 2 - Nominal composition: 1/2 Cr-1/2 Mo) | | | |
| **Low alloy steel** | | | | | |
| A182 | Grade F1 | 1 | 70 | 480 | C-Mo forgings and fittings |
| | Grade F2 | 2 | 70 | 480 | Cr-Mo forgings and fittings |
| A209 | Grade T1 | 1 | 55 | 380 | C-Mo seamless tubes |
| | Grade T1a | 1 | 60 | 415 | |
| | Grade T1b | 1 | 53 | 565 | |
| A213 | Grade T2 | 2 | 60 | 415 | Cr-Mo seamless tubes |
| A250 | Grade T1 | 1 | 55 | 380 | C-Mo welded tubes |
| | Grade T1a | 1 | 60 | 415 | |
| | Grade T1b | 1 | 53 | 565 | |
| A335 | Grade P1 | 1 | 55 | 386 | C-Mo pipe |
| | Grade P2 | 2 | 55 | 386 | Cr-Mo pipe |
| | Grade P15 | 1 | 60 | 415 | C-Si-Mo pipe |
| A369 | Grade FP1 | 1 | 55 | 380 | C-Mo pipe |
| | Grade FP2 | 2 | 55 | 380 | Cr-Mo pipe |
| A426 | Grade CP1 | 1 | 55 | 380 | C-Mo cast pipe |
| | Grade CP2 | 2 | 55 | 380 | Cr-Mo cast pipe |
| | Grade CP15 | 1 | 60 | 415 | C-Si-Mo cast pipe |
| P-number 4 | | (Group 1 - Nominal composition: 1-2 Cr and 1/2 Mo) (Group 2 - Nominal composition: 1/2-1 Ni) | | | |
| **Low alloy steel** | | | | | |
| A182 | Grade F11 | 1 | 70 | 480 | Cr-Mo forgings and fittings |
| | Grade F12 | 1 | 70 | 480 | |
| A199 | Grade T3b | 1 | 60 | 415 | Cr-Mo seamless tubes |
| | Grade T11 | 1 | 60 | 415 | |
| A200 | Grade T3b | 1 | 60 | 415 | Cr-Mo seamless tubes |
| | Grade T11 | 1 | 60 | 415 | |
| A213 | Grade T3b | 1 | 60 | 415 | Cr-Mo seamless tubes |
| | Grade T11 | 1 | 60 | 415 | |
| | Grade T12 | 1 | 60 | 415 | |
| A333 | Grade 4 | 2 | 60 | 415 | 3/4 Cr-3/4 Ni pipe for low temperature service |
| A335 | Grade P11 | 1 | 60 | 415 | 1-1/4 Cr-Mo seamless pipe |
| | Grade P12 | 1 | 60 | 415 | 1 Cr-Mo seamless pipe |

## Table 2
## Grouping of materials for qualification for the AR-1 level (cont'd)

| ASTM material specification | | Group No. | Tensile strength, min, ksi | MPa | Type of material |
|---|---|---|---|---|---|
| P-number 4 (cont'd) | | | | | |
| A369 | Grade FP3b | 1 | 60 | 415 | 2 Cr-Mo forged and bored pipe |
| | Grade FP11 | 1 | 60 | 415 | 1-1/4 Cr-Mo forged and bored pipe |
| | Grade FP12 | 1 | 60 | 415 | 1 Cr-Mo forged and bored pipe |
| A423 | Grade 1 | 2 | 60 | 415 | 3/4 Cr-1/2 Ni tubes |
| | Grade 2 | 2 | 60 | 415 | 3/4 Cr-1/2 Cu tubes |
| A426 | Grade CP11 | 1 | 70 | 480 | 1-1/4 Cr-Mo cast pipe |
| | Grade CP12 | 1 | 60 | 415 | 1 Cr-Mo cast pipe |
| P-number 5 | | (Group 1 - Nominal composition: 2-1/2 Cr-Mo) | | | |
| **Alloy steel** | | (Group 2 - Nominal composition: 5 Cr-Mo) | | | |
| | | (Group 3 - Nominal composition: 7-9 Cr-Mo) | | | |
| A182 | Grade F5 | 2 | 60 | 415 | 5 Cr-Mo forgings and fittings |
| | Grade F5a | 2 | 80 | 550 | 5 Cr-Mo forgings and fittings |
| | Grade F7 | 3 | 60 | 415 | 7 Cr-Mo forgings and fittings |
| | Grade F9 | 3 | 85 | 985 | 9 Cr-Mo forgings and fittings |
| | Grade F21 | 1 | 75 | 515 | 3 Cr-Mo forgings and fittings |
| | Grade F22 | 1 | 75 | 515 | 2-1/4 Cr-Mo forgings and fittings |
| A199 | Grade T4 | 1 | 60 | 415 | 2-1/2 Cr-Mo seamless tubes |
| | Grade T5 | 2 | 60 | 415 | 5 Cr-Mo seamless tubes |
| | Grade T7 | 3 | 60 | 415 | 7 Cr-Mo seamless tubes |
| | Grade T9 | 3 | 60 | 415 | 9 Cr-Mo seamless tubes |
| | Grade T21 | 1 | 60 | 415 | 3 Cr-Mo seamless tubes |
| | Grade T22 | 1 | 60 | 415 | 2-1/4 Cr-Mo seamless tubes |
| A200 | Grade T4 | 1 | 60 | 415 | 2-1/2 Cr-Mo seamless tubing |
| | Grade T21 | 1 | 60 | 415 | 3 Cr-Mo seamless tubing |
| | Grade T22 | 1 | 60 | 415 | 2-1/4 Cr-Mo seamless tubing |
| A213 | Grade T5 | 2 | 60 | 415 | 5 Cr-Mo seamless tubes |
| | Grade T5b | 2 | 60 | 415 | 5 Cr-Mo-Si seamless tubes |
| | Grade T5c | 2 | 60 | 415 | 4 Cr-Mo-Ti seamless tubes |
| | Grade T7 | 3 | 60 | 415 | 7 Cr-Mo seamless tubes |
| | Grade T9 | 3 | 60 | 415 | 9 Cr-Mo seamless tubes |
| | Grade T21 | 1 | 60 | 415 | 3 Cr-Mo seamless tubes |
| | Grade T22 | 1 | 60 | 415 | 2-1/4 Cr-Mo seamless tubes |
| A335 | Grade P5 | 2 | 60 | 415 | 5 Cr-Mo seamless pipe |
| | Grade P5b | 2 | 60 | 415 | 5 Cr-Mo-Si seamless pipe |
| | Grade P5c | 2 | 60 | 415 | 5 Cr-Mo-Ti seamless pipe |
| | Grade P7 | 3 | 60 | 415 | 7 Cr-Mo seamless pipe |
| | Grade P21 | 1 | 60 | 415 | 3 Cr-Mo seamless pipe |
| | Grade P22 | 1 | 60 | 415 | 2-1/4 Cr-Mo seamless pipe |
| | Grade P9 | 3 | 60 | 415 | 9 Cr-Mo seamless pipe |
| A369 | Grade FP21 | 1 | 60 | 415 | 3 Cr-Mo forged and bored pipe |
| | Grade FP5 | 2 | 60 | 415 | 5 Cr-Mo forged and bored pipe |
| | Grade FP7 | 3 | 60 | 415 | 7 Cr-Mo forged and bored pipe |
| | Grade FP9 | 3 | 60 | 415 | 9 Cr-Mo forged and bored pipe |
| | Grade FP22 | 1 | 60 | 415 | 2-1/4 Cr-Mo forged and bored pipe |
| A426 | Grade CP5 | 2 | 90 | 620 | 5 Cr-Mo centrifugally cast pipe |
| | Grade CP7 | 3 | 60 | 415 | 7 Cr-Mo centrifugally cast pipe |
| | Grade CP9 | 3 | 90 | 620 | 9 Cr-Mo centrifugally cast pipe |
| | Grade CP21 | 1 | 60 | 415 | 3 Cr-Mo centrifugally cast pipe |
| | Grade CP22 | 1 | 76 | 515 | 2-1/4 Cr-Mo centrifugally cast pipe |

**Table 2**
**Grouping of materials for qualification for the AR-1 level (cont'd)**

| ASTM material specification | | Tensile strength, min, ksi | MPa | Type of material | |
|---|---|---|---|---|---|
| P-number 6 | | | | | |
| **High alloy steel - predominently martensitic** | | | | | |
| A268 | Grade TP-409 | 60 | 415 | 11 Cr-Ti tubes | |
| | Grade TP-410 | 60 | 415 | 13 Cr tubes | |
| | Grade TP-429 | 65 | 456 | 15 Cr tubes | |
| A426 | Grade CPCA15 | 90 | 620 | 13 Cr centrifugally cast pipe | |
| P-number 7 | | | | | |
| **High alloy steel - predominently ferritic** | | | | | |
| A268 | Grade 405 | 60 | 415 | 12 Cr-A1 tubes | |
| | Grade 430[b] | 60 | 415 | 16 Cr tubes | |
| | Grade XM-8 | 60 | 415 | 18 Cr-Ti tubes | |
| P-number 8 | | | | | |
| **High alloy steel - austenitic** | | | | | |
| A182 | Grade F304 | 75 | 515 | 18 Cr-8 Ni | Forgings and fittings |
| | Grade F304H | 75 | 515 | 18 Cr-8 Ni | |
| | Grade F304L | 65 | 450 | 18 Cr-8 Ni | |
| | Grade 310 | 95 | 655 | 25 Cr-20 Ni-Cb | |
| | Grade F304N | 80 | 550 | 18 Cr-8 Ni-N | |
| | Grade F316 | 75 | 515 | 18 Cr-12 Ni-2 Mo | |
| | Grade 316H | 75 | 515 | 18 Cr-12 Ni-2 Mo | |
| | Grade 316L | 65 | 450 | 18 Cr-12 Ni-2 Mo | |
| | Grade 316N | 80 | 550 | 17 Cr-12 Ni-2 Mo-V | |
| | Grade F321 | 75 | 515 | 18 Cr-10 Ni-Ti | |
| | Grade F321H | 75 | 515 | 18 Cr-10 Ni-Ti | |
| | Grade F347 | 75 | 515 | 18 Cr-10 Ni-Cb | |
| | Grade F347H | 75 | 515 | 18 Cr-10 Ni-Cb | |
| | Grade 348 | 75 | 515 | 18 Cr-10 Ni-Cb | |
| | Grade 348H | 75 | 515 | 18 Cr-10 Ni-Cb | |
| | Grade F10 | 80 | 556 | 20 Ni-8 Cr | |
| A213 | Grade TP-304 | 75 | 515 | 18 Cr-8 Ni | Seamless tubes |
| | Grade TP-304H | 75 | 515 | 18 Cr-8 Ni | |
| | Grade TP-304L | 70 | 480 | 18 Cr-8 Ni | |
| | Grade TP-304N | 80 | 550 | 18 Cr-8 Ni-N | |
| | Grade TP-310 | 75 | 515 | 25 Cr-20 Ni | |
| | Grade TP-316 | 75 | 515 | 18 Cr-12 Ni-2 Mo | |
| | Grade TP-316H | 75 | 515 | 18 Cr-12 Ni-2 Mo | |
| | Grade TP-316L | 70 | 480 | 18 Cr-12 Ni-2 Mo | |
| | Grade TP-316N | 80 | 550 | 16 Cr-12 Ni-2 Mo-N | |
| | Grade TP-321 | 75 | 515 | 18 Cr-10 Ni-Ti | |
| | Grade TP-321H | 75 | 515 | 18 Cr-10 Ni-Ti | |
| | Grade TP-347 | 75 | 515 | 18 Cr-10 Ni-Cb | |
| | Grade TP-347H | 75 | 515 | 18 Cr-10 Ni-Cb | |
| | Grade TP-348 | 75 | 515 | 18 Cr-10 Ni-Cb | |
| | Grade TP-348H | 75 | 515 | 18 Cr-10 Ni-Cb | |
| | Grade XM-15 | 75 | 515 | 18 Cr-18 Ni | |

## Table 2
## Grouping of materials for qualification for the AR-1 level (cont'd)

| ASTM material specification | | Tensile strength, min, ksi | MPa | Type of material | |
|---|---|---|---|---|---|
| P-number 8 (cont'd) | | | | | |
| **High alloy steel - austenitic** | | | | | |
| A249 | Grade TP-304 | 75 | 515 | 18 Cr-8 Ni | Welded tubes |
| | Grade TP-304H | 75 | 515 | 18 Cr-8 Ni | |
| | Grade TP-304L | 70 | 480 | 18 Cr-8 Ni | |
| | Grade TP-304N | 80 | 550 | 18 Cr-8 Ni-N | |
| | Grade TP-309 | 75 | 515 | 23 Cr-12 Ni | |
| | Grade TP-310 | 75 | 515 | 25 Cr-20 Ni | |
| | Grade TP-316 | 75 | 515 | 18 Cr-12 Ni-2 Mo | |
| | Grade TP-316H | 75 | 515 | 18 Cr-12 Ni-2 Mo | |
| | Grade TP-316L | 70 | 480 | 18 Cr-12 Ni-2 Mo | |
| | Grade TP-316N | 80 | 550 | 16 Cr-12 Ni-2 Mo-N | |
| | Grade TP-317 | 75 | 515 | 19 Cr-13 Ni-3 Mo | |
| | Grade TP-321 | 75 | 515 | 18 Cr-10 Ni-Ti | |
| | Grade TP-321H | 75 | 515 | 18 Cr-10 Ni-Ti | |
| | Grade TP-347 | 75 | 515 | 18 Cr-10 Ni-Cb | |
| | Grade TP-347H | 75 | 515 | 18 Cr-10 Ni-Cb | |
| | Grade TP-348 | 75 | 515 | 18 Cr-10 Ni-Cb | |
| | Grade TP-348H | 75 | 515 | 18 Cr-10 Ni-Cb | |
| | Grade XM-15 | 75 | 515 | 18 Cr-18 Ni | |
| A271 | Grade TP-304 | 75 | 515 | 18 Cr-8 Ni | Seamless tubing |
| | Grade TP-304H | 75 | 515 | 18 Cr-8 Ni | |
| | Grade TP-321 | 75 | 515 | 18 Cr-10 Ni-Ti | |
| | Grade TP-321H | 75 | 515 | 18 Cr-10 Ni-Ti | |
| | Grade TP-347 | 75 | 515 | 18 Cr-10 Ni-Cb | |
| | Grade TP-347H | 75 | 515 | 18 Cr-10 Ni-Cb | |
| A312 | Grade TP-304 | 75 | 515 | 18 Cr-8 Ni | Pipe |
| | Grade TP-304H | 75 | 515 | 18 Cr-8 Ni | |
| | Grade TP-304L | 70 | 480 | 18 Cr-8 Ni | |
| | Grade TP-304N | 80 | 550 | 18 Cr-8 Ni-N | |
| | Grade TP-309 | 75 | 515 | 23 Cr-12 Ni-Cb | |
| | Grade TP-310 | 75 | 515 | 25 Cr-20 Ni-Cb | |
| | Grade TP-316 | 75 | 515 | 18 Cr-12 Ni-2 Mo | |
| | Grade TP-316H | 75 | 515 | 18 Cr-12 Ni-2 Mo | |
| | Grade TP-316L | 70 | 480 | 18 Cr-12 Ni-2 Mo | |
| | Grade TP-316N | 80 | 550 | 16 Cr-12 Ni-2 Mo-N | |
| | Grade TP-317 | 75 | 515 | 19 Cr-13 Ni-3 Mo | |
| | Grade TP-321 | 75 | 515 | 18 Cr-10 Ni-Ti | |
| | Grade TP-321H | 75 | 515 | 18 Cr-10 Ni-Ti | |
| | Grade TP-347 | 75 | 515 | 18 Cr-10 Ni-Cb | |
| | Grade TP-347H | 75 | 515 | 18 Cr-10 Ni-Cb | |
| | Grade TP-348 | 75 | 515 | 18 Cr-10 Ni-Cb | |
| | Grade TP-348H | 75 | 515 | 18 Cr-10 Ni-Cb | |
| | Grade TP-XM15 | 75 | 515 | 18 Cr-18 Ni | |
| A376 | Grade TP-340 | 75 | 515 | 18 Cr-8 Ni | Seamless pipe |
| | Grade TP-304H | 75 | 515 | 18 Cr-8 Ni | |
| | Grade TP-304N | 80 | 550 | 18 Cr-8 Ni-N | |
| | Grade TP-316 | 75 | 515 | 18 Cr-12 Ni-2 Mo | |
| | Grade TP-316H | 75 | 515 | 18 Cr-12 Ni-2 Mo | |
| | Grade TP-316N | 80 | 550 | 16 Cr-12 Ni-2 Mo-N | |
| | Grade TP-321 | 75 | 515 | 18 Cr-10 Ni-Ti | |
| | Grade TP-321H | 75 | 515 | 18 Cr-10 Ni-Ti | |
| | Grade TP-347 | 75 | 515 | 18 Cr-10 Ni-Cb | |
| | Grade TP-347H | 75 | 515 | 18 Cr-10 Ni-Cb | |
| | Grade TP-348 | 75 | 515 | 18 Cr-10 Ni-Cb | |

**Table 2**
**Grouping of materials for qualification for the AR-1 level (cont'd)**

| ASTM material specification | | Tensile strength, min, ksi | MPa | Type of material | |
|---|---|---|---|---|---|
| P-number 8 (cont'd) | | | | | |
| **High alloy steel – austenitic** | | | | | |
| A409 | Grade TP-304 | 75 | 515 | 18 Cr-8 Ni | Welded pipe |
| | Grade TP-304L | 70 | 480 | 18 Cr-8 Ni | |
| | Grade TP-309 | 75 | 515 | 23 Cr-12 Ni | |
| | Grade TP-310 | 75 | 515 | 25 Cr-20 Ni | |
| | Grade TP-316 | 75 | 515 | 18 Cr-12 Ni-2 Mo | |
| | Grade TP-316L | 70 | 480 | 18 Cr-12 Ni-2 Mo | |
| | Grade TP-317 | 75 | 515 | 19 Cr-13 Ni-3 Mo | |
| | Grade TP-321 | 75 | 515 | 18 Cr-10 Ni-Ti | |
| | Grade TP-347 | 75 | 515 | 18 Cr-10 Ni-Cb | |
| | Grade TP-348 | 75 | 515 | 18 Cr-10 Ni-Cb | |
| A430 | Grade FP-304 | 70 | 480 | 18 Cr-8 Ni | Forged and bored pipe |
| | Grade FP-304H | 70 | 480 | 18 Cr-8 Ni | |
| | Grade FP-304N | 80 | 559 | 18 Cr-8 Ni-N | |
| | Grade FP-316 | 70 | 480 | 18 Cr-12 Ni-2 Mo | |
| | Grade FP-316H | 70 | 480 | 18 Cr-12 Ni-2 Mo | |
| | Grade FP-316N | 80 | 550 | 16 Cr-12 Ni-2 Mo-N | |
| | Grade FP-321 | 70 | 480 | 18 Cr-10 Ni-Ti | |
| | Grade FP-321H | 70 | 480 | 18 Cr-10 Ni-Ti | |
| | Grade FP-347 | 70 | 480 | 18 Cr-10 Ni-Cb | |
| | Grade FP-347H | 70 | 480 | 18 Cr-10 Ni-Cb | |
| | Grade FP-16-8-2H | 70 | 480 | 16 Cr-8 Ni-2 Mo | |
| A451 | Grade CPF-3 | 70 | 480 | 19 Cr-10 Ni | Centrifugally cast pipe |
| | Grade CPF-3A | 77 | 530 | 19 Cr-10 Ni | |
| | Grade CPF-3M | 70 | 480 | 19 Cr-11 Ni-2-1/2 Mo | |
| | Grade CPF-8 | 77 | 530 | 18 Cr-8 Ni | |
| | Grade CPF-8A | 70 | 480 | 18 Cr-8 Ni | |
| | Grade CPF-8M | 70 | 480 | 18 Cr-9 Ni-2 Mo | |
| | Grade CPF-10MC | 70 | 480 | 17 Cr-14 Ni-2 Mo-Cb | |
| | Grade CPH-10 | 70 | 480 | 25 Cr-12 Ni | |
| | Grade CPF-8C | 70 | 480 | 18 Cr-9 Ni-Cb | |
| | Grade CPH-8 | 65 | 450 | 25 Cr-12 Ni | |
| | Grade CPK-20 | 65 | 450 | 25 Cr-12 Ni | |
| | Grade CPH-20 | 70 | 480 | 25 Cr-12 Ni | |
| A452 | Type TP-304H | 75 | 515 | 18 Cr-8 Ni | Centrifugally cast and cold worked pipe |
| | Type TP-316H | 75 | 515 | 18 Cr-12 Ni-2 Mo | |
| | Type TP-347H | 75 | 515 | 18 Cr-10 Ni-Cb | |
| A688 | Grade TP-304 | 75 | 515 | 18 Cr-8 Ni | Welded pipe |
| | Grade TP-304L | 70 | 480 | 18 Cr-8 Ni | |
| | Grade TP-316 | 75 | 515 | 18 Cr-12 Ni-Mo | |
| | Grade TP-316L | 70 | 480 | 18 Cr-12 Ni-Mo | |
| | Grade TP-304N | 80 | 550 | 18 Cr-8 Ni | |
| | Grade TP-316N | 80 | 550 | 18 Cr-12 Ni-2 Mo | |
| P-number 9A[c] | (Nominal composition: 2 to 2-1/2 nickel steel) | | | | |
| **Nickel alloy steel** | | | | | |
| A33 | Grade 7 | 65 | 450 | 2-1/2 Ni | Pipe for low temperature service |
| | Grade 9 | 63 | 435 | 2 Ni-1 Cu | |
| A334 | Grade 7 | 65 | 450 | 2-1/2 Ni | Tubes for low temperature service |
| | Grade 9 | 63 | 435 | 2 Ni-1 Cu | |

**Table 2**
**Grouping of materials for qualification for the AR-1 level (cont'd)**

| ASTM material specification | | Tensile strength, min, ksi | MPa | Type of material |
|---|---|---|---|---|
| P-number 9B | | (Nominal composition: 3-4 nickel steel) | | |
| **Nickel alloy steel** | | | | |
| A333 | Grade 3 | 65 | 456 | 3-1/2 Ni pipe for low temperature service |
| A334 | Grade 3 | 65 | 456 | 3-1/2 Ni tubes for low temperature service |
| P-number 10B[b] | | | | |
| **Specialty high alloy steels** | | | | |
| A213 | Grade T17 | 60 | 415 | 1 Cr-V seamless tubes |
| P-number 10E[b] | | | | |
| **Specialty high alloy steels** | | | | |
| A268 | Grade TP-446 | 70 | 480 | 27 Cr } Tubing |
| A268 | Grade TP-329 | 90 | 620 | 26 Cr-4 Ni-Mo } Tubing |
| P-number 10H[b] | | | | |
| **Specialty high alloy steels** | | | | |
| A669 | | 92 | 634 | 18 Cr-5 Ni-3 Mo Seamless tubes |
| P-number 10I[b] | | | | |
| **Specialty high alloy steels** | | | | |
| A182 | Grade FXM-27 | 60 | 415 | 26 Cr-1 Mo forgings and fittings |
| A268 | Grade XM-27 | 65 | 450 | 26 Cr-1 Mo } Tubing |
| A268 | Grade XM-33 | 68 | 470 | 26 Cr-1 Mo-Ti } Tubing |
| A731 | Grade XM-27 | 65 | 450 | 26 Cr-1 Mo } Pipe |
| A731 | Grade XM-33 | 65 | 450 | 26 Cr-1 Mo-Ti } Pipe |
| P-number 11 | | (Nominal composition: 8-9 nickel steel) | | |
| **Nickel alloy steel** | | | | |
| A333 } A334 } | Grade 8 | 100 | 690 | 9 Ni pipe |

**Table 2**
**Grouping of materials for qualification for the AR-1 level (cont'd)**

| ASTM material specification | | | Tensile strength, min, ksi | MPa | Type of material |
|---|---|---|---|---|---|
| **P-number 21** | | | | | |
| **Aluminum and aluminum-base alloys** | | | | | |
| | | Temper | | | |
| B241, B345 | 1060 | All | 8.5 | 60 | Seamless tubular products |
| B210, B234 | 1060 | All | 9 | 62 | Seamless tubular products |
| B234 | 3003 | All | 14 | 96 | Seamless tubular products |
| B241, B345 | Alclad 3003 | All | 13 | 90 | Seamless tubular products |
| **P-number 22** | | | | | |
| **Aluminum-base alloys** | | | | | |
| B210 | 5052 | All | 25 | 72 | Seamless tubular products |
| B234 | 5154 | All | 30 | 206 | Seamless tubular products |
| B241 | 5454 | All | 31 | 214 | Seamless tubular products |
| **P-number 23** | | | | | |
| **Aluminum-base alloys** | | | | | |
| B210, B234 | 6061 | -T4, -T6 | 24 | 165 | Seamless tubular products |
| B241, B345 | 6063 | -T5, -T6 | 17 | 117 | Seamless tubular products |
| **P-number 25** | | | | | |
| **Aluminum-base alloys** | | | | | |
| B210 | 5083 | All | 39 | 269 | Seamless tubular products |
| B241, B345 | 5456 | All | 41 | 283 | Seamless tubular products |
| B345 | 5086 | All | 35 | 241 | Seamless tubular products |
| **P-number 31** | | | | | |
| **Copper and copper alloy** | | | | | |
| | Alloy designation | | | | |
| B42 | C10200, C12000, C12200 | | 30 | 206 | Seamless copper pipe |
| B75 | C10200, C12000, C12200, C14200 | | 30 | 206 | Seamless copper pipe |
| B88 | C10200, C12000, C12200 | | 30 | 206 | Seamless copper tube |

### Table 2
### Grouping of materials for qualification for the AR-1 level (cont'd)

| ASTM material specification | Alloy designation | Tensile strength, min, ksi | MPa | Type of material |
|---|---|---|---|---|
| P-number 31 (cont'd) | | | | |
| **Copper and copper alloy** | | | | |
| B111 | C10200, C12000, C12200, C14200 | 30 | 206 | Copper and copper alloy seamless tube |
| B447 | C10200, C12000, C12200, C14200 | 30 | 206 | Welded copper tube |
| B543 | C12200 | 32 | 220 | Welded copper tube |
| B586 | C19400 | 45 | 310 | Welded copper and copper alloy tube |
| P-number 35 | | | | |
| **Copper-base alloy** | | | | |
| B111 | C60800 | 50 | 345 | Copper and copper alloy seamless tube |
| B395 | C60800 | 50 | 345 | Copper and copper alloy seamless tube |
| P-number 41* | | | | |
| **Nickel** (99% Ni) | | | | |
| B161 | N02200 | 55 | 380 | Seamless pipe and tube |
| | N02201 | 50 | 345 | Seamless pipe and tube, low carbon |
| B163 | N02200 | 55 | 380 | Seamless tube |
| | N02201 | 50 | 345 | Seamless tube, low carbon |
| P-number 42* | | | | |
| **Nickel-base alloy** (≃ 67 Ni-30 Cu) | | | | |
| B163 | N04400 | 70 | 480 | Seamless tube |
| B165 | N04400 | 70 | 480 | Seamless pipe and tube |
| P-number 43* | | | | |
| **Nickel-chromium-iron alloy** | | | | |
| B163 | N06600 | 80 | 550 | Seamless tube (72Ni-15Cr-8Fe) |
| B167 | N06600 | 80 | 550 | Seamless pipe and tube (72Ni-15Cr-8Fe) |
| B444 | N06625 | 120 | 830 | Seamless pipe and tube (60Ni-22Cr-9Mo-4Cb) |
| B619 | N06002 | 100 | 690 | Welded pipe (48Ni-22Cr-9Mo-18Fe) |
| B622 | N06002 | 100 | 690 | Seamless pipe and tube (48Ni-22Cr-9Mo-18Fe) |
| B626 | N06002 | 100 | 690 | Welded tube (48Ni-22Cr-9Mo-18Fe) |

**Table 2**
**Grouping of materials for qualification for the AR-1 level (cont'd)**

| ASTM material specification | Alloy designation | Tensile strength, min, ksi | Tensile strength, min, MPa | | Type of material |
|---|---|---|---|---|---|
| **P-number 44** | | | | | |
| **Nickel-molybdenum and nickel-molybdenum-chromium alloys** | | | | | |
| B619 | N06455 | 100 | 690 | Welded pipe | (65Ni-16Mo-16Cr-low carbon) |
| | N10001 | 110 | 760 | | (65Ni-28Mo-5Fe) |
| | N10276 | 100 | 690 | | (60Ni-16Mo-16Cr-Fe-W-low carbon) |
| | N10665 | 100 | 690 | | (70Ni-28Mo) |
| B622 | N06455 | 100 | 690 | Seamless pipe and tubing | (65Ni-16Mo-16Cr-low carbon) |
| | N10001 | 100 | 690 | | (65Ni-28Mo-5Fe) |
| | N10276 | 100 | 690 | | (60Ni-16Mo-16Cr-Fe-W-low carbon) |
| | N10665 | 110 | 760 | | (70Ni-28Mo) |
| **P-number 45** | | | | | |
| **Nickel-chromium-iron alloys** | | | | | |
| B163 | N08800 | 75 | 515 | Seamless tube | (33Ni-21Cr) |
| | N08810 | 65 | 450 | | (33Ni-21Cr) |
| | N08825 | 85 | 585 | | (42Ni-22Cr-3Mo-2Cu) |
| B407 | N08800 | 75 | 515 | Seamless pipe and tube | (33Ni-21Cr) |
| | N08810 | 65 | 450 | | (33Ni-21Cr) |
| B423 | N08825 | 75 | 515 | Seamless pipe and tube | (42Ni-22Cr-3Mo-2Cu) |
| B464 | N08020 | 85 | 585 | Seamless and welded pipe | (36Ni-20Cr-2Mo-3Cu-Cb) |
| B468 | N08020 | 85 | 585 | Seamless and welded tube | (36Ni-20Cr-2Mo-3Cu-Cb) |
| B619 | N06007 | 90 | 620 | Welded pipe | (44Ni-22Cr-6Mo-Cu-Cb) |
| | N06975 | 85 | 585 | | (50Ni-24Cr-6Mo-Cu-Ti) |
| | N08320 | 75 | 515 | | (26Ni-22Cr-5Mo-Ti) |
| B622 | N06007 | 90 | 620 | Seamless pipe and tube | (44Ni-22Cr-6Mo-Cu-Cb) |
| | N06975 | 85 | 585 | | (50Ni-24Cr-6Mo-Cu-Ti) |
| | N08320 | 75 | 515 | | (26Ni-22Cr-5Mo-Ti) |
| **P-number 46** | | | | | |
| **Nickel-chromium-silicon-iron alloys** | | | | | |
| B535 | N08330 | 70 | 480 | Seamless and welded pipe | (35Ni-19Cr) |

**Table 2**
**Grouping fo materials for qualification for the AR-1 level (cont'd)**

| ASTM material specification | | Tensile strength, min, ksi | MPa | Type of material |
|---|---|---|---|---|
| P-number 51 | | | | |
| B337 | Grade 1 | 35 | 220 | Unalloyed titanium pipe |
| | Grade 2 | 50 | 345 | |
| | Grade 7 | 50 | 345 | |
| B338 | Grade 1 | 35 | 220 | Unalloyed titanium pipe |
| | Grade 2 | 50 | 345 | |
| | Grade 7 | 50 | 345 | |
| P-number 52 | | | | |
| B337 | Grade 3 | 65 | 450 | Unalloyed titanium pipe |
| B338 | Grade 3 | 65 | 450 | Unalloyed titanium pipe |
| P-number 61 | | | | |
| B523 | Grade 702 | 52 | 358 | Unalloyed Zr tubing |

a. The materials within P-number 1 have been divided into three groups based on fracture toughness, and this division should be disregarded unless impact tests are performed on the sample test welds. If fracture toughness is a consideration, qualification for Group 3 qualifies for Groups 2 and 1, but not vice versa. Similarly, qualification for Group 2 qualifies for Group 1, but not vice versa.
b. Materials listed in P-number groups 10B, 10E, 10H, and 10I require separate qualification.
c. Qualification with P-number 9B qualifies for P-number 9A, but not vice versa.

*Higher minimum tensile strength is shown in some specifications if material is not in annealed condition or for small diameter or thin walled forms.

2.3.9 **Welding Position.** The addition of any welding position other than those already qualified shall require requalification (also see paragraph 1.4).

2.3.10 **Direction of Welding.** When the pipe is in the 5G or 6G position, a change in the direction of welding (that is, from bottom to top or top to bottom) shall require requalification.

2.3.11 **Preheat Temperature.** A change of 100° F (55° C) or more in the specified preheat temperature range shall require requalification.

2.3.12 **Interpass Temperature.** A change in the minimum or maximum specified interpass temperature shall require requalification.

2.3.13 **Postheating**

(1) For ferrous materials, a change of 100° F (55° C) or more in the specified postheating temperature shall require requalification.

(2) A change in the maximum temperature that takes the material above a phase transformation temperature (such as the lower critical temperature for steel) shall require requalification.

# 3. AR-1 Level of Welder Qualification

*Editor's Note:* The AR-2 level of welding procedure and welder qualification described in the 1969 edition of this standard has been incorporated into the AR-1 level in this revised edition (see paragraph 1.1).

## 3.1 General Notes on Welder Qualification for Groove Welds

Refer to Table 6 for sample welds and tests required for AR-1 welder qualification. Prepare only those sample welds required to qualify the welder for the pipe or tube diameter, wall thickness ranges, and positions of welding that will be involved in production welding.

A welder who prepares the sample welds for the successful qualification of a welding procedure is qualified to weld with that procedure without further qualification.

**Table 3**
**A-number classification of ferrous filler metal by chemical composition for the AR-1 level**

| Filler metal composition number | Type of weld deposit | Analysis, percent (Single values shown are max) C | Cr | Mo | Ni | Mn | Si | AWS Filler metal specification and classification number | Reference P-number for pipe |
|---|---|---|---|---|---|---|---|---|---|
| A-1 | Mild steel | 0.15 | | | | 2.25 | 1.00 | A5.1, A5.2, A5.17, A5.18, A5.20 | P-1 |
| A-2 | Carbon-1/2%moly | 0.15 | | 0.40-0.65 | | 2.10 | 1.00 | A5.5 - Class E70XX-A1; A5.23 - Class EA1, EA2, and EA3; A5.28 - Class ER80S-D2 | P-1 and P-3 |
| A-3 | Chrome (1/4-2%)-1/2%moly | 0.15 | 0.50-2.00 | 0.40-0.65 | | 1.50 | 1.00 | A5.5 - Class E80XX-B1 and E80XX-B2; A5.23 - Class EB1, EB2, and EB2H; A5.28 - Class ER80S-B2, ER80S-B2L, E80C-B2, E80C-B2L | P-3 and P-4 |
| A-4 | Chrome (2-6%)-moly (1/2-1-1/2%) | 0.15 | 2.00-6.00 | 0.40-1.50 | | 1.00 | 2.00 | A5.5 - Class E90XX-B3, E90XX-B4; A5.23 - Class EB3, EB5, EB6, and EB6H; A5.28 - Class ER90S-B3, ER90S-B3L, E90C-B3, E90C-B3L | P-4 and P-5 |
| A-5 | Chrome (6-10%)-moly (1/2-1-1/2%) | 0.15 | 6.00-10.00 | 0.40-1.50 | | 1.00 | 2.00 | A5.4 - Class E502, E505, and E7Cr; A5.9 - Class ER502; A5.22 - Class E502T-X, E505T-X | P-5 |
| A-6 | High alloy chrome-martensitic | 0.15 | 11.00-15.00 | 0.70 | | 2.00 | 1.00 | A5.4 - Class E410; A5.9 - Class ER410, ER420; A5.22 - Class E410T-X | P-6 |
| A-7 | High alloy chrome-ferritic | 0.15 | 11.00-30.00 | 1.00 | | 1.00 | 3.00 | A5.4 - Class E430; A5.9 - Class ER430; A5.22 - Class E430T-X | P-7 |
| A-8 | High alloy chromium-nickel (these grades are usually partially ferritic) | 0.15 | 14.50-24.00 | 4.00 | 7.50-15.00 | 2.50 | 1.00 | A5.4 - Class E308, E308L, E309, E309Cb, E309Mo, E312, E16-8-2, E316, E316L, E317, E318, E347, and E349; A5.9 - Class ER308, ER308L, ER309, ER312, ER347, ER348, and ER349; A5.22 - Class E308T-X, E308LT-X, E308MoT-X, E308MoLT-X, E309T-X, E316T-X, E316LT-X, and E347T-X | P-8 |
| A-9 | High alloy chromium-nickel (fully austenitic) | 0.30 | 24.00-28.00 | 4.00 | 15.00-37.00 | 2.50 | 1.00 | A5.4 - Class E310, E310Cb, E310Mo, E330; A5.9 - Class ER310 | P-8 |
| A-10 | Nickel steel (to 4%) | 0.15 | | 0.55 | 0.8-4.00 | 1.50 | 1.00 | A5.28 - Class ER80S-NiX, E80C-NiX; A5.5 - Class E80XX-CX; A5.23 - Class ENi-X | P-9 |
| A-11 | Manganese-moly steel | 0.17 | | 0.25-0.75 | 0.85 | 1.25-2.25 | 1.00 | A5.5 - Class E90XX-DX, E100XX-DX; A5.23 - Class ED1 and ED2 | P-3 and P-11 |

**Table 4**
**F-number grouping of welding rods and electrodes for carbon, alloy, and stainless steels for the AR-1 level**

| AWS Specification no. | F1 | F2 | F3 | F4 | F5 | F6 |
|---|---|---|---|---|---|---|
| A5.1 | EXX20<br>EXX24<br>EXX27<br>EXX28 | EXX12<br>EXX13<br>EXX14 | EXX10<br>EXX11 | EXX15<br>EXX16<br>EXX18 | | |
| A5.5 | EXX20<br>EXX27 | EXX13 | EXX10<br>EXX11 | EXX15<br>EXX16<br>EXX18 | | |
| A5.4 | | | | EXXX15<br>EXXX16 | EXXX15<br>EXXX16 | |
| A5.2 | | | | | | RXX |
| A5.9 | | | | | | ERXXX |
| A5.17 | | | | | | FXX-EXXXX |
| A5.18 | | | | | | ERXXS-X |
| A5.20 | | | | | | EXXT-X |
| A5.22 | | | | | | EXXXT-X |
| A5.23 | | | | | | FXX-EXXX-X<br>FXX-ECXXX-X<br>FXX-EXXX-XN<br>FXX-ECXXX-XN |
| A5.28 | | | | | | ERXXS-X<br>EXXC-X |

To determine if a welder's existing qualifications will permit him to weld with a new welding procedure, refer to paragraph 3.3, "Limitations on Changes Allowable in Welding Variables Controlling the Qualification of a Welder to the AR-1 Level." See Appendix A for a discussion of the possibility of qualifying a welder to two or more welding procedures through the performance of a single qualification test series.

The required welder qualification tests are divided into a nondestructive and a destructive test series. Welders may be qualified by use of either the nondestructive or the destructive series of required tests within the following restrictions:

(1) Welders may be qualified by either the nondestructive or the destructive series of required tests for base metals falling into P-number groupings 1 through 11, 31 through 35, and 41 through 45 with all welding processes or combinations of welding processes

(2) When using the gas tungsten arc welding process, welders may be qualified by either the nondestructive or the destructive series of required tests for base metals falling into P-number groupings 21 through 25 and 51 and 52.

(3) Welders must be qualified by the destructive series of required tests for all other combinations of base metal groupings and welding processes.

The nondestructive series of required tests utilizes visual inspection and radiography. The destructive series of required tests utilizes a sectional examination for small diameter pipe under 1-1/2 in. (38 mm) nominal pipe size and the guided-bend test for all larger pipe sizes.

The pipe coupons for the sample test welds shall be made of the same P-number materials as that specified in the welding procedure. However, for welder qualification on ferrous materials falling within the P-number groupings 1 through 8, carbon steel P-number 1 coupons may be substituted for the P-number material specified in the welding procedure, provided that the filler metal used for the welder test welds is identical to that specified by the welding procedure. For welder qualification on aluminum alloys falling into the P-numbers 21 through 25 groupings, pipe coupons of any material from the P-numbers 21 through 25 groupings may be substituted for the material specified in the welding procedure, provided the filler metal specified in the welding procedure is used.

**Table 5**
**F-number grouping of nonferrous electrodes and welding rods for the AR-1 level**

| F-number | AWS Specification and classification number | | | |
|---|---|---|---|---|
| | Aluminum and aluminum alloys | Copper and copper alloys | | |
| | A5.10 | A5.6 | A5.7 | A5.27 |
| F-21 | ER1100 | | | |
| F-22 | ER5183<br>ER5356<br>ER5554<br>ER5556<br>ER5654 | | | |
| F-23 | ER4043<br>ER4047 | | | |
| F-31 | | ECu | ERCu | |
| F-32 | | ECuSi | ERCuSi | |
| F-33 | | ECuSn-A<br>ECuSn-C | ERCuSn-A | |
| F-34 | | ECuNi | ERCuNi | |
| F-35 | | | | RBCuZn-A<br>RCuZn-C |
| F-36 | | ECuA1-A2<br>ECuA1-B | ERCuA1-A1<br>ERCuA1-A2<br>ERCuA1-A3 | |
| F-37 | | ECuNiA1<br>ECuMnNiA1 | ERCuNiA1<br>ERCuMnNiA1 | |

The welder qualification test may be terminated at any stage of the testing whenever it becomes apparent to the supervisor conducting the test that the welder is not following the welding procedures or does not have the skill required to produce satisfactory results. One such indication is the need for an excessive number of repairs to produce an acceptable weld.

**3.2 Tests Required for Sample Groove Welds.** Details regarding specimen preparation, testing techniques, and acceptance requirements for qualification of a welder to the AR-1 level follow.

**3.2.1 Visual Inspection.** Each sample weld shall be inspected visually with the unaided eye:

(1) *Prior to assembly* of the test pieces, to assure that the proper materials are being used, that the parts have been cleaned, and that the end preparation is correct.

(2) *After fit-up,* to check the joint clearances.

(3) *During welding,* to inspect for cracks, porosity, undercut, incomplete fusion, surface roughness, and poor crater conditions. Grinding may be used when necessary on starts and stops and to remove local defects from bead-to-bead, but dressing of the completed sample weld shall be limited to the removal of local defects appearing on the outer surface of the weld reinforcement. If production welding is to be accomplished from the outside of the pipe only, no repair of the root of the sample weld from the inside of the pipe is permitted.

(4) *After welding has been completed* (including any postheating), to detect any surface defects.

The following are the acceptance requirements for a sample weld to pass visual inspection:

(a) The specified base metal and filler metal shall have been used.

(b) The weld shall be free of cracks.

(c) All craters shall be filled to the full cross section of the weld and shall be free of cracks upon completion of the welding.

**Table 5**
**F-number grouping of nonferrous electrodes and welding rods for the AR-1 level (cont'd)**

| F-number | AWS Specification and classification number | | | |
|---|---|---|---|---|
| | Nickel and nickel alloys | | Titanium and titanium alloys | Zirconium and zirconium alloys |
| | A5.11 | A5.14 | A5.16 | A5.24 |
| F-41 | ENi-1 | ERNi-1 | | |
| F-42 | ENiCu-7 | ERNiCu-7 | | |
| F-43 | ERNiCrFe-1<br>ERNiCrFe-2<br>ERNiCrFe-3<br>ENiCrFe-4<br>ENiCrMo-2<br>ENiCrMo-3 | ERNiCr-3<br>ERNiCrFe-5<br>ERNiCrFe-7<br>ERNiCrMo-2<br>ERNiCrMo-3 | | |
| F-44 | ENiMo-1<br>ENiMo-3<br>ENiCrMo-4<br>ENiCrMo-5 | ERNiMo-1<br>ERNiMo-2<br>ERNiMo-3<br>ERNiMo-7<br>(Alloy B-2)<br>ERNiCrMo-4<br>ERNiCrMo-5<br>ERNiCrMo-7<br>(Alloy C-4) | | |
| F-45 | ENiCrMo-1 | ERNiCrMo-1<br>ERNiFeCr-1 | | |
| F-51 | | | ERTi-1<br>ERTi-2<br>ERTi-3<br>ERTi-4 | |
| F-61 | | | | ERZr-1<br>ERZr-2<br>ERZr-3 |

(d) The face of the weld shall be at least flush with the outside surface of the pipe, and the weld shall merge smoothly with the base metal. Depth of the weld undercut shall not exceed 1/32 in. (0.8 mm), irrespective of length. Weld reinforcement shall not exceed the following:

| Sample weld pipe size, in. | mm | Face reinforcement, max, in. | mm |
|---|---|---|---|
| 1/2 | 13 | 3/32 | 2.4 |
| 2 or 3 | 51-76 | 1/8 | 3.2 |
| 4 and over | 102 | 3/16 | 4.8 |

(e) The root of the weld shall be examined, and there shall be complete joint penetration and no cracks, nor incomplete fusion. A concave root surface is permitted within the limits shown in (f) below, provided the total weld thickness is equal to or greater than that of the base metal.

(f) A concave root surface and root reinforcement are permitted within the limits shown below:

**Limits of surface imperfections on the inside surface of the joint**

| Sample weld pipe size, in. | mm | Maximum root surface concavity, in. | mm | Maximum root reinforcement, in. | mm |
|---|---|---|---|---|---|
| 1/2 | 13 | 1/32 | 0.8 | 1/16 | 1.6 |
| 2 or 3 | 51-76 | 1/16 | 1.6 | 1/16 | 1.6 |
| 4 and over ≤ 0.75 in. wall | 102 | 1/16 | 1.6 | 3/32 | 2.4 |
| 6 and over > 0.75 in. wall | 152 | 1/16 | 1.6 | 1/8 | 3.2 |

## Table 6
## Sample welds and tests required for AR-1 level of welder qualification

| Sample welds | | | | | Tests required on sample welds[a,b] | | | | | | | | | |
|---|---|---|---|---|---|---|---|---|---|---|---|---|---|---|
| | | | | | Nondestructive test series | | | Destructive test series | | | | | | |
| | | | | | | | | | Bend tests[g] Number of specimens, | | | | | |
| | | | | | | | | | All positions except 5G & 6G | | | 5G and 6G positions only | | |
| Pipe size of sample welds[c] | Pipe or tube wall thicknesses qualified, min, in. | Pipe or tube wall thicknesses qualified, max, in. | Number of sample welds per position | Pipe or tube size qualified, NPS, in. | Visual inspection[d] | Radiography[e] | Visual inspection[d] | Sectional[f] examination, no. of specimen edges examined[f] | Face | Root | Side | Face | Root | Side |
| | | | | | Wall thicknesses 0.750 in. and under | | | | | | | | | |
| 1/2 in. Sch. 40 | 0.065 | 0.400 | 1 | Through 1-1/2 | Yes | Yes | Yes | 4 | 0 | 0 | 0 | 0 | 0 | 0 |
| 2 in. Sch. 80 or 3 in. Sch. 40 | 0.065 | 0.674 | 1 | 1 through 4 | Yes | Yes | Yes | 0 | 1 | 1 | 0 | 1 | 1 | 0 |
| 6 in. Sch. 40 | 0.109 | 0.750 | 1 | Over 4 | Yes | Yes | Yes | 0 | 1 | 1 | 0 | 2 | 2 | 0 |
| | | | | | Wall thicknesses over 0.750 in. | | | | | | | | | |
| 6, 8, or 10 in. Sch. 120 | 0.375 | Any | 1 | 6 and over | Yes | Yes | Yes | 0 | 0 | 0 | 2 | 0 | 0 | 4 |

a. Welders may be qualified using either the destructive or the nondestructive series of tests. See 3.1 for limitations.
b. Carbon steel pipe coupons may be substituted for more expensive alloys for sample welds in some materials. See 3.1 for allowable substitutions.
c. See Appendix C for option of job size pipe.
d. See 3.2.1.
e. See 3.2.2.
f. See 3.2.3 and 3.2.4.
g. See 3.2.3 and 3.2.5. Side-bends shall be used instead of face- and root-bends for sample weld wall thicknesses of more than 0.375 in.

SI Equivalents

| in. | mm | in. | mm | in. | mm | in. | mm | in. | mm |
|---|---|---|---|---|---|---|---|---|---|
| 0.065 | 1.65 | 0.400 | 10.16 | 1/2 | 13 | 2 | 51 | 6 | 152 |
| 0.109 | 2.77 | 0.674 | 17.12 | 1 | 26 | 3 | 76 | 8 | 203 |
| 0.375 | 9.52 | 0.750 | 19.05 | 1-1/2 | 38 | 4 | 102 | 10 | 252 |

**3.2.2 Radiographic Examination.** Radiographic examination shall not be used to locate sound or defective areas upon which to perform thereafter destructive tests in order to qualify or disqualify a welder. The preparation of sample welds, general and specific radiographic technique details, and acceptance requirements for welder qualification shall be the same as described in paragraph 2.2.3 for welding procedure qualification.

**3.2.3 Preparation of Sample Welds for Destructive Tests**

**3.2.3.1 1/2 in. (13 mm) Sample Welds.** Prepare the sample weld for examination by sectioning it along the axis of the pipe to expose four cross sections of the weld for examination, as shown in Fig. 3.2.3.1.

**3.2.3.2 2 or 3 in. (51 or 76 mm) Sample Welds.** For all positions, remove and prepare one face- and one root-bend specimen from the sample weld as shown in Fig. 3.2.3.2.

**3.2.3.3 6, 8, and 10 in. (152, 203, and 254 mm) Sample Welds.** For the 1G and 2G positions, remove and prepare one face-and one root-bend specimen as shown in Fig. 3.2.3.3. For wall thicknesses over 3/8 in. (9.5 mm), two side-bend specimens may be substituted for the face-and root-bend specimens.

For the 5G and 6G positions, remove and prepare two face- and two root-bend specimens as shown in Fig. 3.2.3.3. For wall thicknesses of 3/8 in. (9.5 mm), four side-bend specimens may be substituted for the two face- and two root-bend specimens.

**3.2.4 Sectional Examination.** Polish and etch the four edges of the sample weld exposed by the sectioning to a contrast suitable to reveal any defects that may be present in the weld. Etchants that may be used are listed in Appendix D.

The polished and etched sample weld cross sections shall be examined, and specimens exhibiting any of the following defects shall be judged unacceptable:

(1) Incomplete fusion between the weld metal and the base metal.

(2) Incomplete fusion between any of the weld passes.

(3) Lack of complete joint penetration.

(4) When evaluating slag inclusions and porosity, the sum of the maximum dimensions of each slag inclusion and pore in a given cross section shall not exceed 1/3 of the weld thickness.

**3.2.5 Guided-Bend Test.** The guided-bend test for welder qualification shall be performed in the same manner as for welding procedure qualification, as outlined in paragraph 2.2.6.

In order to meet the acceptance requirements for the test, the bent specimens shall have no cracks or other open defects exceeding 1/8 in. (3.2 mm) measured in any direction on the convex surface of the specimen, except that cracks occurring at the corners shall not be considered unless there is definite evidence that they result from slag inclusions or other internal defects.

*Note:* Details for optional bend tests are described in Appendix C.

**3.3 Limitations on Changes Allowable in Welding Variables Controlling the Qualification of a Welder to the AR-1 Level**

Information concerning all of the following applicable welding variables shall be recorded in the welder qualification test record. A welder qualified to one welding procedure is also qualified to weld with other qualified welding procedures, provided that the welding variables of such procedures do not vary from those of the procedure to which the welder was qualified by more than the limits for such changes listed below. If the welding variables of the production welding procedure to be used vary from those in the welder's qualification test record by more than the limits listed below, the welder shall be requalified to the new procedure.

**3.3.1 Filler Metal**

**3.3.1.1 Composition.** For ferrous materials, a change in the chemical composition from one A-number to any other A-number shown in Table 3 shall require requalification, except that qualification of a welder with filler metal A-numbers 1 through 5 automatically qualifies him for that A-number, all lower A-numbers, and one A-number higher than the one tested. A-numbers 7, 8, and 9 require individual qualification.

**3.3.1.2 Usability.** For all except aluminum alloys, a change from one F-number to another F-number in Tables 4 or 5, or to any other filler metal, shall require requalification, except that qualification of a welder with F-number groupings 1 through 4 automatically qualifies him for that F-number grouping and all lower grouping filler metals.

For aluminum alloys, qualification with filler metal in any F-number 2X grouping shall qualify the welder to weld with any other F-number 2X filler metal.

**3.3.2 Pipe Size and Wall Thickness.** A change from one diameter wall thickness combination to another, as shown in Table 6, shall require requalification.

For the oxyacetylene or oxyhydrogen process, the maximum thickness qualified is the thickness of the sample test weld.

**3.3.3 Backing** (for complete joint penetration groove welds only)

(1) In any welding process, the addition or omission of a preplaced consumable insert shall require requalification.

(2) The omission of a backing ring in joints welded from one side only shall require requalification.

(3) For gas tungsten arc welding, the omission of filler metal for the root pass shall require requalification.

(4) For open butt joints, the omission of a gas backing shall require requalification.

(5) The omission of a backing weld shall require requalification.

(6) For the oxyacetylene process, the addition of backing shall require requalification.

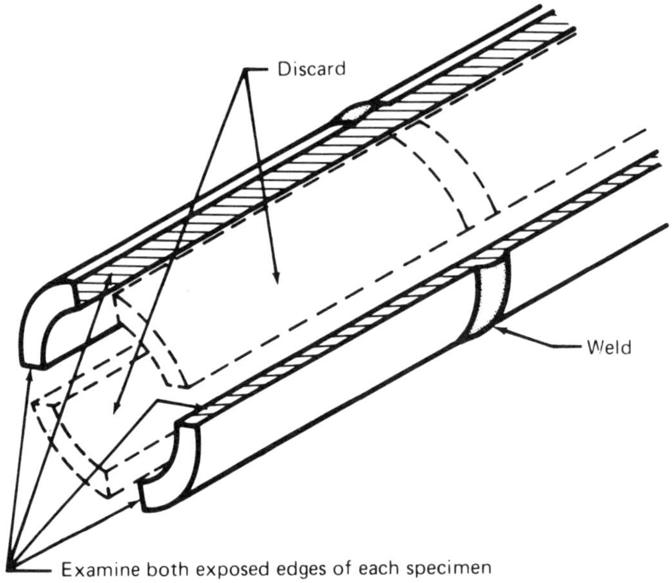


**Fig. 3.2.3.1—Preparation of sample weld for examination**

**3.3.4 Welding Process.** A change from one welding process to any other welding process, or combination of processes, shall require requalification. Where a combination of welding processes is required, each welder must be qualified for the particular process or processes that he will be required to perform in production welding.

**3.3.5 Gas Shielding.** In gas metal arc, gas tungsten arc, or flux cored arc welding, a change in the composition of the gas involving one or more of the following shall require requalification:

(1) A change from one gas to another.

(2) The addition or omission of an active gas such as oxygen, hydrogen, or carbon dioxide.

(3) The addition or omission of more than 5 percent by volume of another inert gas or carbon dioxide.

(4) With gas mixtures, a 5 percent change in the volume of one of the gases.

(5) A decrease of 10 percent or more in the gas flow rate.

**3.3.6 Electrical Characteristics**

(1) A change in the welding current from AC to DC, or vice versa, shall require requalification.

(2) A change in polarity shall require requalification.

(3) For the gas metal arc process, a change in current density that results in a change from spray arc, globular arc, or pulsating arc to a short circuiting arc, or vice versa, shall require requalification.

**3.3.7 Welding Position.** The addition of any welding position other than those already qualified shall require requalification (see paragraph 1.4).

**3.3.8 Direction of Welding.** When the pipe is in the 5G or 6G position, a change in the direction of welding, that is from bottom to top or top to bottom, shall require requalification.

# 4. AR-3 Level of Welding Procedure and Welder Qualification

### 4.1 General Notes on Welding Procedure and Welder Qualification for Groove Welds

Refer to Table 7 for sample welds and tests required for both the AR-3 welding procedure qualification and the AR-3 welder qualification. Prepare only those sample welds necessary to qualify the pipe diameter-wall thickness ranges and the positions of welding that will be involved in production welding.

If it becomes apparent that the welding conditions set forth in the tentative welding procedure will not produce a joint that meets the requirements, the test should be discontinued. One such indication is the need for an excessive number of repairs to produce an acceptable weld. In such a case, new conditions must be established and their validity proven by a new test weld.

Table 7 also gives sample weld sizes and tests required for AR-3 welder qualification testing. The tests required and acceptance requirements for welding procedure and welder qualification for the AR-3 level are identical.

A welder who prepares the sample welds for the successful qualification of a welding procedure is qualified to weld with that procedure without further testing.

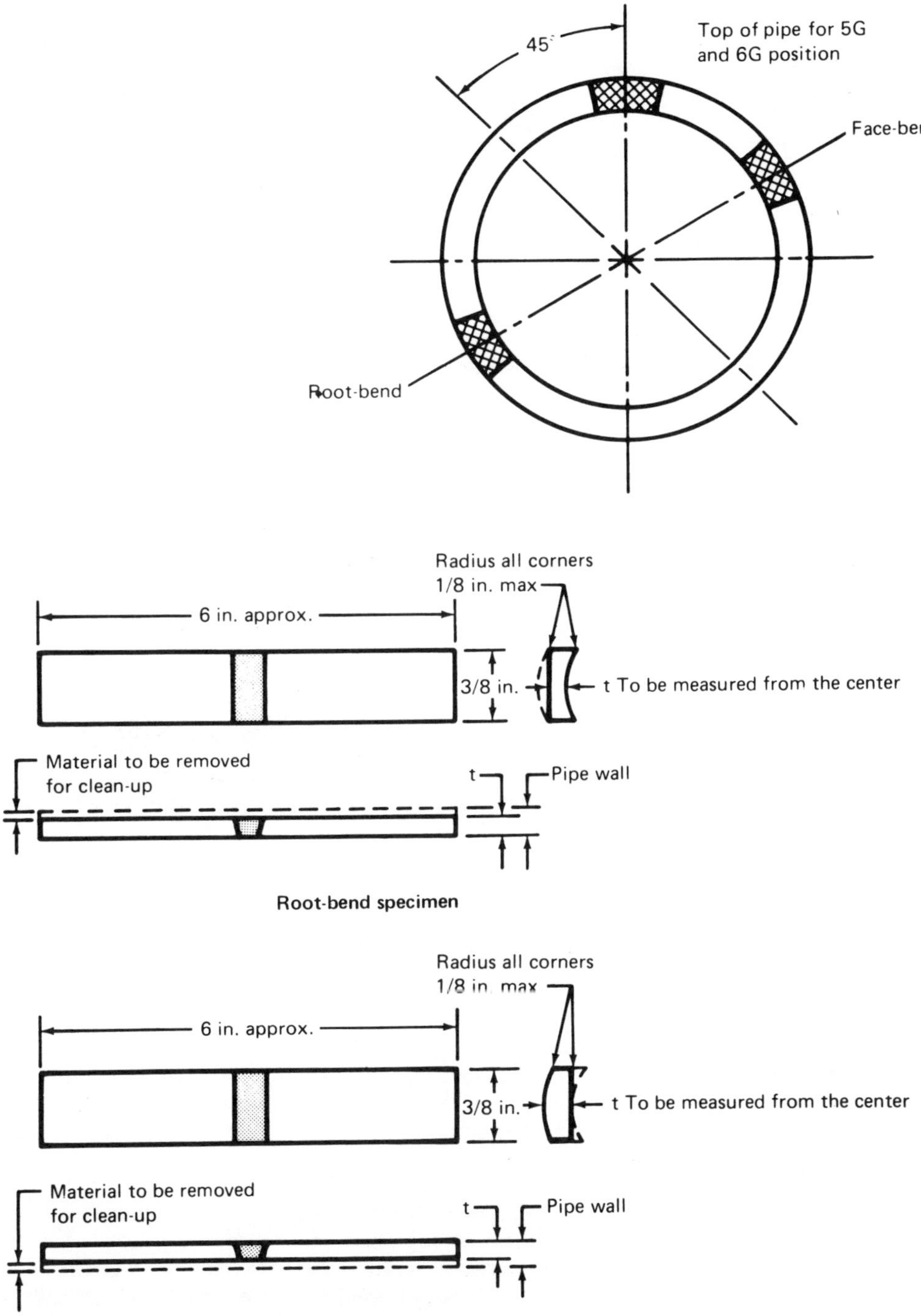


SI Equivalents

| in. | mm |
|---|---|
| 1/8 | 3.2 |
| 3/8 | 9.5 |
| 6 | 152 |

Notes:

1. All dimensions in inches.
2. Weld reinforcement and backing, if any, shall be removed flush with the surface of the specimen. Cut surfaces shall be smooth and parallel.

**Fig. 3.2.3.2–Preparation of face-and root-bend specimens for 2 or 3 in. sample welds**

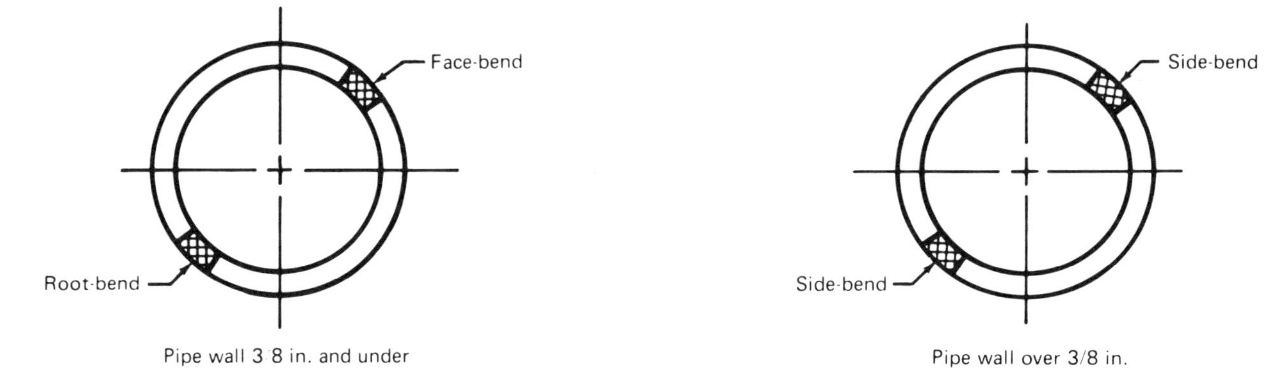


**Location of specimens for 1G and 2G positions**

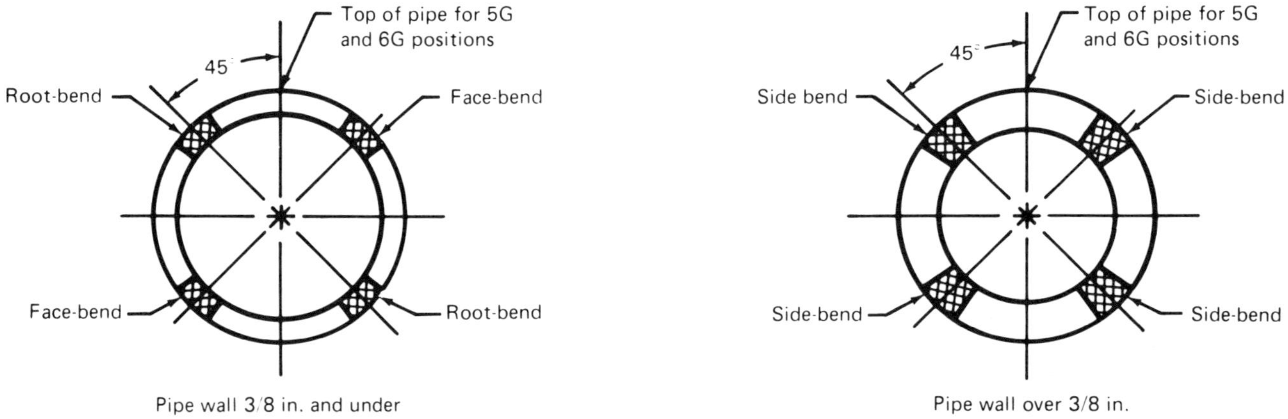


**Location of specimens for 5G and 6G positions**

**Fig. 3.2.3.3—Preparation of face- and root-bend specimens for 6, 8, and 10 in. sample welds**

Paragraph 4.3 should be referred to prior to the start of the procedure qualification testing, to determine all of the variables that need to be controlled and documented. The limits on allowable changes in welding variables controlling the qualification of a welding procedure for the AR-3 level in paragraph 4.3 should be referred to in order to determine when modification of an existing welding procedure requires the performance of new qualification tests.

To determine if a welder's existing qualifications will permit him to weld with a new welding procedure, refer to paragraph 4.4, "Limitations on Changes Allowable in Welding Variables Controlling the Qualification of a Welder to the AR-3 Level." See Appendix A for a discussion of the possibility of qualifying a welder to two or more welding procedures through the performance of a single qualification test series.

The welder qualification test may be terminated at any stage of the testing whenever it becomes apparent to the supervisor conducting the test that the welder is not following the welding procedure or does not have the skill required to produce satisfactory results. One such indication is the need for an excessive number of repairs to produce an acceptable weld.

**4.2 Tests Required for Sample Groove Welds.** Details regarding specimen preparation, testing techniques, and acceptance requirements for qualification of a welding procedure or a welder, or both, to the AR-3 level follow.

**4.2.1 Visual Inspection.** Each sample weld shall be inspected visually with the unaided eye.

(1) *Prior to assembly* of the test pieces, to assure that the proper materials are being used, that the parts have been cleaned, and that the end preparation is correct.

(2) *After fit-up,* to check the joint clearances.

(3) *During welding,* to inspect for cracks, porosity, undercut, incomplete fusion, surface roughness, and poor crater conditions. Each weld bead should be inspected as-deposited and after any normal mechanical operations

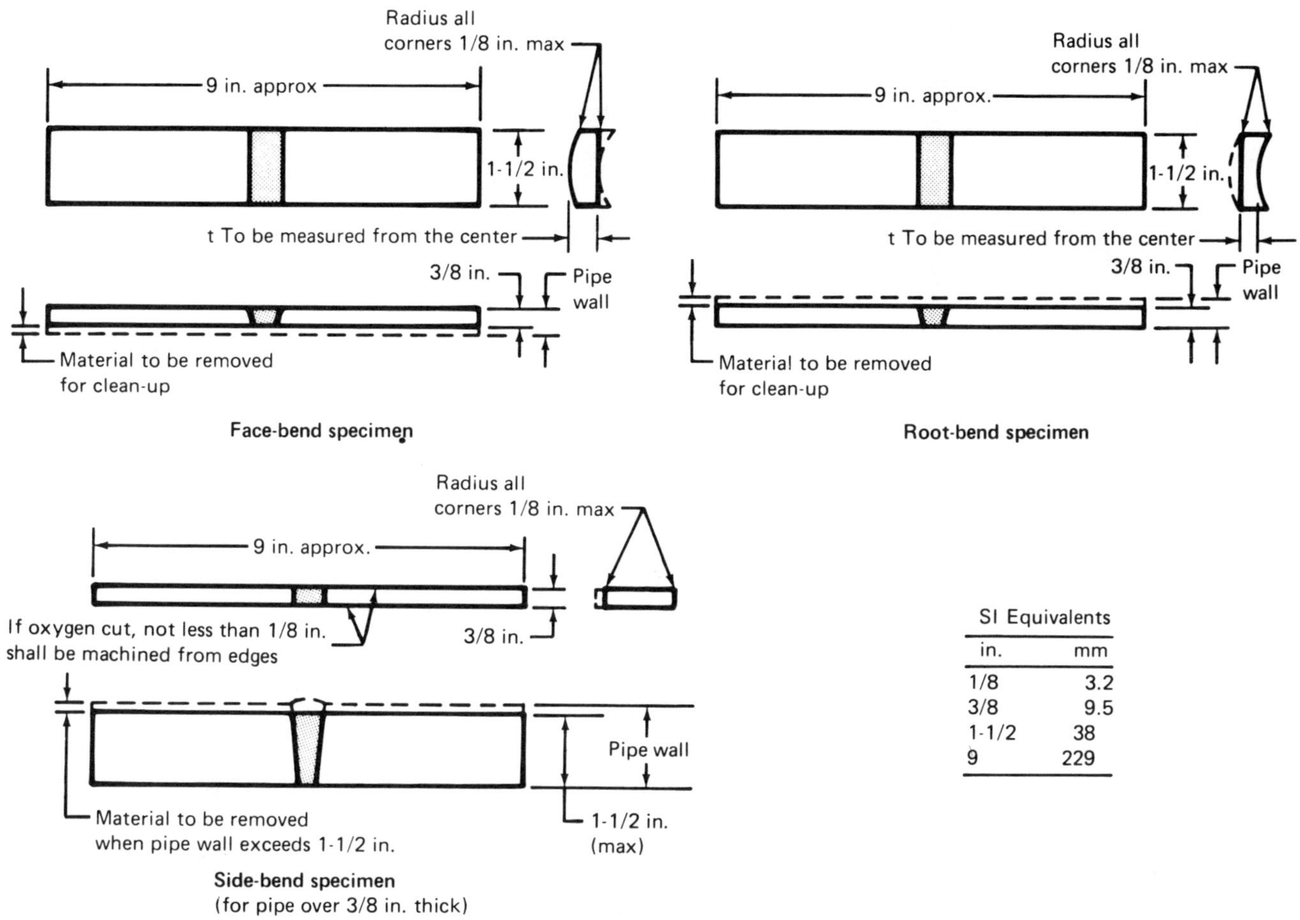


SI Equivalents

| in. | mm |
|---|---|
| 1/8 | 3.2 |
| 3/8 | 9.5 |
| 1-1/2 | 38 |
| 9 | 229 |

Notes:

1. All dimensions in inches.
2. Weld reinforcement and backing, if any, shall be removed flush with the surface of the specimen. Cut surfaces shall be smooth and parallel.

**Fig. 3.2.3.3 (continued)—Preparation of face- and root-bend specimens for 6, 8, and 10 in. sample welds**

such as brushing, filing, or grinding, prior to depositing subsequent passes. Grinding may be used when necessary on starts and stops and to remove local defects from bead-to-bead, but dressing of the complete sample weld shall be limited to the removal of weld reinforcement. If production welding is to be accomplished from the outside of the pipe only, no repair of the root of the sample weld from the inside of the pipe is permitted.

(4) *After welding has been completed,* to detect any surface defects.

Following are the acceptance requirements for a sample weld to pass visual inspection:

(a) The specified base metal and filler metal shall have been used.

(b) The weld shall be free of cracks.

(c) All craters shall be filled to the full cross section of the weld, and shall be free of cracks upon completion of the welding.

(d) The face of the weld shall be at least flush with the outside surface of the pipe, and the weld shall merge smoothly with the base metal. Depth of weld undercut shall not exceed 1/32 in. (0.8 mm), irrespective of length. Weld reinforcement shall not exceed the following:

| Sample weld pipe size, in. | mm | Face reinforcement, in. | mm |
|---|---|---|---|
| Through 1-1/2 | 38 | 3/32 | 2.4 |
| 2 or 3 | 51-76 | 1/8 | 3.2 |
| 4 and over | 102 | 3/16 | 4.8 |

(e) The root of the weld shall be examined, and there shall be complete joint penetration and no cracks, nor incomplete fusion. A concave root surface is permitted within the limits shown in (f) below, provided the total weld thickness is equal to or greater than that of the base metal.

**Table 7**
**Sample welds and tests required for AR-3 level of welding procedure and welder qualification**

| Pipe size of sample welds | Pipe or tube wall thicknesses qualified, min, in. | max, in.[a] | Number of sample welds per position | Pipe or tube size qualified, NPS, in. | Visual[b] inspection | Pressure bend,[c] number of tests | Macrostructure[d] examination, number of specimen edges examined | Alternate[e] bend test, number of specimens: Face | Root |
|---|---|---|---|---|---|---|---|---|---|
| 1/2, 3/4, 1, 1-1/4 or 1-1/2 in. Sch. 40 | 0.065 | 0.154 | 1 | Through 1-1/2 | Yes | 1 | 0 | 0 | 0 |
| 2, 4, 6, 8, or 10 in. Sch. 40 | 0.154 | 0.375 | 1 | 2 through 20 | Yes | 0 | 4 | 1 | 1 |

a. Prepare only one size sample weld from among the suggested sizes for each diameter-wall thickness range to be qualified.
b. See 4.2.1.
c. See 4.2.2.
d. See 4.2.3.
e. See 4.2.4. The guided-bend test may be used in lieu of the macrostructure examination.

SI Equivalents

| in. | mm | in. | mm | in. | mm |
|---|---|---|---|---|---|
| 0.065 | 1.6 | 1 | 25 | 6 | 152 |
| 0.154 | 3.9 | 1-1/4 | 32 | 8 | 203 |
| 0.375 | 9.5 | 1-1/2 | 38 | 10 | 252 |
| 1/2 | 13 | 2 | 51 | 20 | 508 |
| 3/4 | 19 | 4 | 102 | | |

(f) A concave root surface and root reinforcement are permitted within the limits shown below.

**Limits of surface imperfections on the inside surface of the joint**

| Sample weld pipe size, in. | mm | Maximum root surface concavity, in. | mm | Maximum root reinforcement, in. | mm |
|---|---|---|---|---|---|
| Through 1-1/2 | 38 | 1/32 | 0.8 | 1/16 | 1.6 |
| 2 and over | 51 | 1/16 | 1.6 | 1/8 | 3.2 |

**4.2.2 Pressure Bend Test.** Prepare the sample weld chosen from the 1/2 in. (13 mm) to 1-1/2 in. (38 mm) size range for pressure bending by welding caps on its ends as shown in Fig. 4.2.2 for all materials except aluminum alloy, P23. For P23, use the macrostructure examination described in paragraph 4.2.3.

Fill with water at city pressure (25 psig minimum [172 MPag]), and close the valve. Pressure bend around a 12 in. (305 mm) diameter mandrel as shown in Fig. 4.2.2.

The weld meets the acceptance requirements if there are no leaks after a 30 degree bend.

**4.2.3 Macrostructure Examinations.** Section, polish, and etch sample welds 2 in. (51 mm) in diameter and over, as shown in Fig. 4.2.3.

In order to meet the acceptance requirements, the etched cross sections shall show complete joint penetration and no cracks, nor incomplete fusion. Root surface concavity shall not exceed 1/16 in. (1.6 mm) in depth. The cumulative slag and porosity shall not exceed 1/8 in. (3.2 mm) in length from the root to the face of the weld.

**4.2.4 Guided-Bend Test.** In lieu of the macrostructure examination described in paragraph 4.2.3, the contractor may perform the following modified guided-bend test. Remove and prepare from the sample weld one face- and one root-bend specimen, as shown in Fig. 4.2.4.

The guided-bend test for welding procedure and welder qualification for the AR-3 level shall be performed in the same manner as for welding procedure qualification for the AR-1 level, as outlined in paragraph 2.2.6.

The sample weld meets the acceptance requirements if none of the specimens undergo complete fracture.

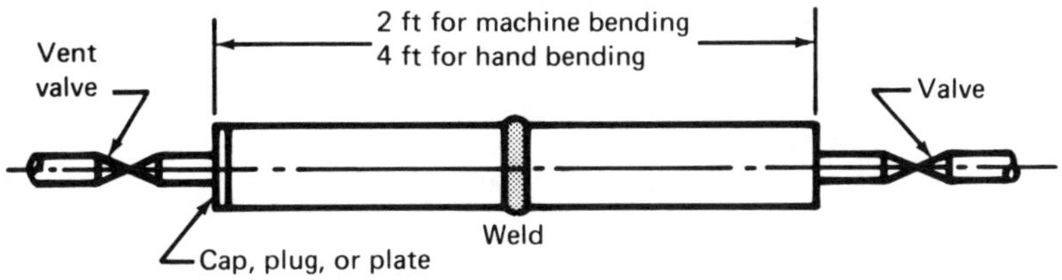


Sample weld

| SI Equivalents | |
|---|---|
| ft | m |
| 2 | 0.61 |
| 4 | 1.22 |

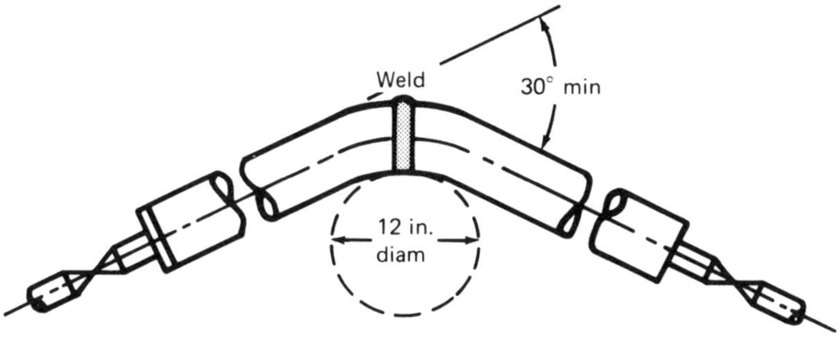


Pressure bending

Fig. 4.2.2–Pressure bend test

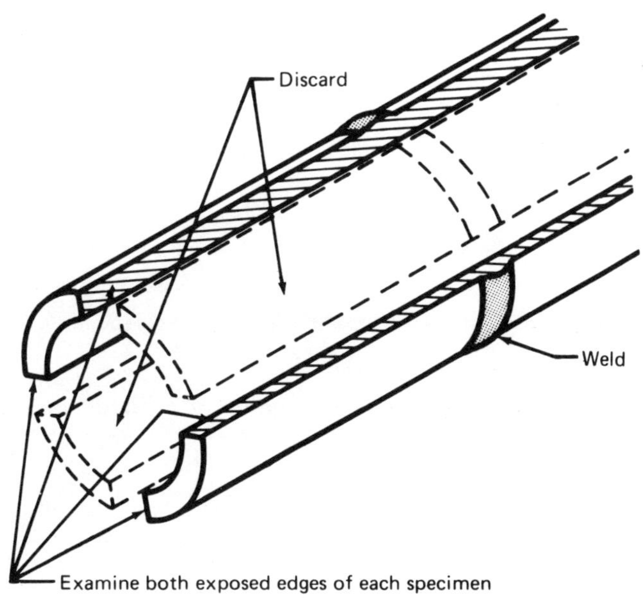


Fig. 4.2.3–Preparation of sample weld for examination

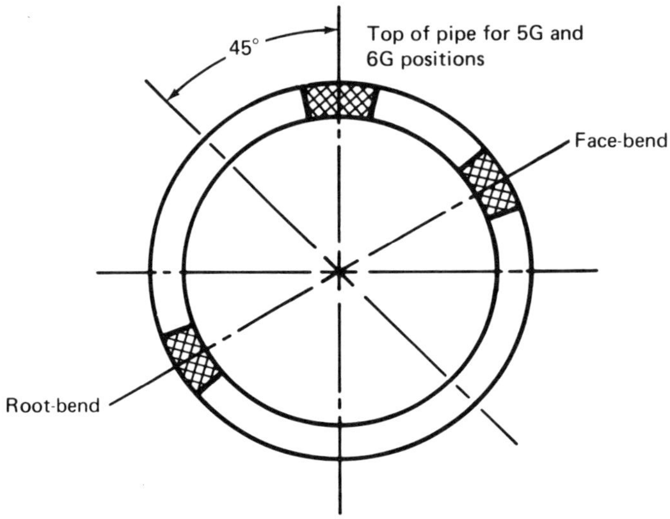


**Location of specimens**

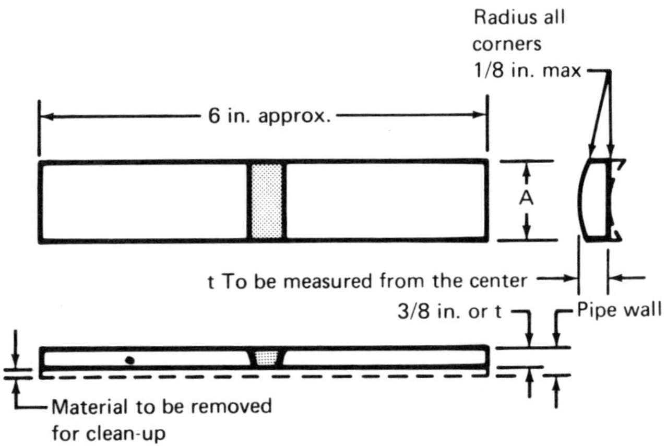


**Face-bend specimen**

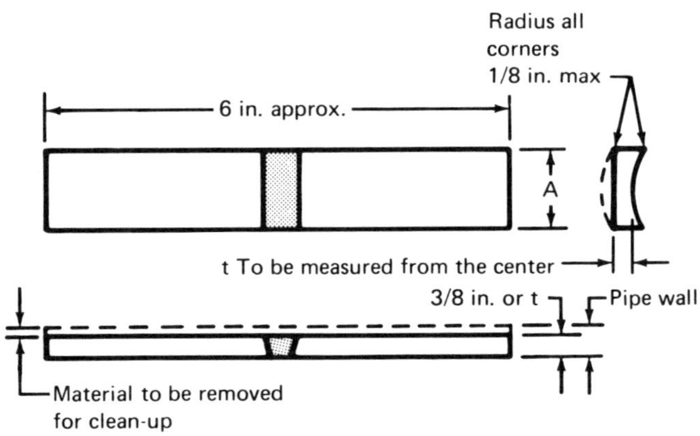


**Root-bend specimen**

Width of face- and root-bend specimens

| Sample weld pipe size, in. | Dimension A, in. |
|---|---|
| 2 through 4 | 3/8 |
| Over 4 | 5/8 |

SI Equivalents

| in. | mm |
|---|---|
| 1/8 | 3.2 |
| 3/8 | 9.5 |
| 5/8 | 15.9 |
| 2-4 | 51-102 |
| 6 | 152 |

Notes:
1. All dimensions in inches.
2. Weld reinforcement and backing, if any, shall be removed flush with the surface of the specimen. Cut surfaces shall be smooth and parallel.

**Fig. 4.2.4—Preparation of specimens for guided-bend testing**

**4.3 Limitations on Changes Allowable in Welding Variables Controlling the Qualification of a Welding Procedure to the AR-3 Level**

Information concerning all of the following applicable welding variables shall be recorded when a welding procedure is first qualified. Changes, other than those given below, may be made without the necessity for qualifying a new welding procedure, provided the existing welding procedure specification is revised to show these changes. The welding procedure shall be requalified and set up as a new procedure when any of the changes listed below are made in the welding procedure.

**4.3.1 Base Metal.** The grouping of materials by P-number classification is based on chemical and metallurgical characteristics. The grouping of filler metals by F-number classification is based on usability characteristics. These groupings are intended to reduce the number of welding procedure qualifications required.

(1) A change from a base metal listed under one P-number in Table 8 to material listed under another P-number, or to any other base metal, shall require requalification.

(2) A change from a similar metal joint to a dissimilar metal joint (two different P-numbers), even though welding procedures have been developed for each of the metals welded to itself, shall require requalification.

**4.3.2 Pipe Size and Wall Thickness.** A change from one diameter wall grouping to another, as shown in Table 7, shall require requalification.

**4.3.3 Joint Design.** A change from a single-V-groove or a single-U-groove to a single-bevel-groove, single-J-groove, or square-groove; or from a single-bevel-groove or a single-J-groove to a square-groove, shall require requalification.

An increase or decrease of 1/16 in. (1.6 mm) or more in root opening and root face dimensions beyond those specified in the welding procedure shall require requalification.

**4.3.4 Backing** (for complete joint penetration groove welds only)

(1) The omission of a backing ring for welds made from one side only shall require requalification.

(2) The addition or omission of a preplaced consumable insert shall require requalification.

(3) The omission of a removable material as a backing device shall require requalification.

(4) The addition or omission of a gas or flux as a means of backing shall require requalification.

(5) The omission of a backing weld shall require requalification.

**4.3.5 Filler Metal**

**4.3.5.1 Composition.** Filler metal or deposited weld metal compositions, other than those listed in Table 9, shall require separate qualification.

**4.3.5.2 Usability.** A change from one F-number to another F-number in Table 10 or 11 shall require requalification.

**4.3.6 Welding Process.** A change from one welding process to another for making either the entire weld or a part thereof shall require requalification. Within a given process, a change from fully automatic to semiautomatic or manual, or a change from semiautomatic to manual, or vice versa, shall require requalification.

Additional requirements or exceptions for various welding processes are listed below:

**4.3.6.1 Oxyacetylene or Oxyhydrogen Process**

(1) The maximum thickness qualified using the oxyacetylene or oxyhydrogen welding process for pipe 4 in. (102 mm) and under is the thickness of the sample weld plus 20 percent.

(2) For pipe sizes over 4 in. (102 mm), the maximum pipe size and wall thickness qualified is the sample weld pipe size and wall thickness plus 20 percent.

(3) The addition of backing rings or backing material shall require requalification.

**4.3.6.2 Gas Tungsten Arc Process**

(1) The addition or omission of filler metal shall require requalification.

(2) The omission of high frequency stabilization shall require requalification.

(3) For single-pass welding, the maximum thickness welded in production shall not exceed the thickness of the test pipe.

**4.3.6.3 Gas Metal Arc Process**

(1) For single-pass welding, the maximum thickness welded in production shall not exceed the thickness of the test pipe.

(2) A change from multipass to single-pass welding shall require requalification, except for single-pass fillet welds not exceeding 1/4 in. (6.4 mm), which do not require requalification.

(3) A change in the filler metal from bare solid electrodes to metal cored electrodes, or vice versa, shall require requalification.

**4.3.6.4 Flux Cored Arc Process**

(1) For single-pass welding, the maximum thickness welded in production shall not exceed the thickness of the test pipe.

(2) A change from multipass to single-pass welding shall require requalification, except for single-pass fillet welds not exceeding 1/4 in. (64 mm), which do not require requalification.

**4.3.6.5 Submerged Arc Process**

(1) A change in the nominal composition of flux shall require requalification.

(2) Any change that would result in the important alloying elements in the weld metal being outside the composition range given in the welding procedure shall require requalification.

(3) A change from single arc to multiple arc, or vice versa, or a change from one type of multiple arc to another type, such as from series arc to parallel arc, shall require requalification.

## Table 8
## Grouping of Materials for Qualification for the AR-3 Level

The grouping of materials in this Table of P-Number classification is based on the similarity of chemical and metallurgical characteristics in order to reduce the number of welding procedure qualifications. While most tubing and piping materials have been grouped below, only a limited number of other product forms have been included. Materials not listed below (including such items as piping, forgings, castings, and plates) may be grouped for qualification purposes with the piping and tubing materials having comparable technical, mechanical and metallurgical characteristics.

| ASTM Material specification | | Tensile strength, min, ksi | MPa | Type of material |
|---|---|---|---|---|
| **P-number 1 Carbon steel** | | | | |
| A53 | Openhearth | 45 | 310 | Furnace welded pipe |
| | Grade A | 48 | 330 | Seamless or welded pipe |
| | Grade B | 60 | 415 | Seamless or welded pipe |
| A105 | | 70 | 480 | Pipe flanges |
| A106 | Grade A | 48 | 330 | Seamless pipe |
| | Grade B | 60 | 415 | |
| | Grade C | 70 | 480 | |
| A120 | | 45 | 310 | Galvanized pipe |
| A135 | Grade A | 48 | 330 | Electric resistance welded pipe |
| | Grade B | 60 | 415 | |
| A139 | Grade A | 48 | 330 | Fusion welded pipe |
| | Grade B | 60 | 415 | |
| | Grade C | 60 | 415 | |
| | Grade D | 60 | 415 | |
| | Grade E | 60 | 415 | |
| A161 | Low carbon | 47 | 325 | Low carbon tubes |
| | Grade T1 | 55 | 380 | C-Mo steel tubes |
| A181 | Grade I | 60 | 415 | Forgings and fittings |
| | Grade II | 70 | 480 | |
| A210 | Grade A1 | 60 | 415 | Seamless tubes |
| | Grade C | 70 | 480 | |
| A226 | | 47 | 325 | Electric resistance welded tubes |
| A500 | Grade A | 45 | 310 | Structural tubing |
| | Grade B | 58 | 400 | |
| | Grade C | 62 | 430 | |
| A501 | | 58 | 400 | Structural tubing |
| A523 | Grade A | 48 | 300 | Conduit pipe |
| | Grade B | 60 | 415 | |
| A524 | Grade I | 60 | 415 | Seamless pipe |
| | Grade II | 55 | 380 | |
| A539 | | 45 | 310 | Coiled tubing |

## Table 8
## Grouping of materials for qualification for the AR-3 Level (cont'd)

| ASTM Material specification | | Temper | Tensile strength, min, ksi | MPa | Type of material |
|---|---|---|---|---|---|
| A556 | Grade A2 | | 47 | 325 | Seamless tubes |
| | Grade B2 | | 60 | 415 | |
| | Grade C2 | | 70 | 480 | |
| A557 | Grade A2 | | 47 | 325 | Electric resistance welded tubes |
| | Grade B2 | | 60 | 415 | |
| | Grade C2 | | 70 | 480 | |
| A587 | | | 48 | 300 | Low carbon pipe |
| A595 | Grade A | | 65 | 450 | Low carbon tapered structural tubing |
| | Grade B | | 70 | 480 | |
| | Grade C | | 70 | 480 | |
| A618 | Grade I | | 70 | 480 | High-strength low alloy structural tubing |
| | Grade II | | 70 | 480 | |
| | Grade III | | 65 | 450 | |
| API 5L | Grade A | | 48 | 300 | Line pipe |
| | Grade B | | 60 | 415 | |
| API 5LS | Grade A | | 48 | 300 | Spiral weld line pipe |
| | Grade B | | 60 | 415 | |
| | Grade X42 | | 60 | 415 | |
| | Grade X46 | | 63 | 435 | |
| | Grade X52 | | 66 | 455 | |
| API 5LX | Grade X42 | | 60 | 415 | High-test line pipe |
| | Grade X46 | | 63 | 435 | |
| | Grade X52 | | 66 | 455 | |
| **P-number 21 Aluminum and aluminum-base alloys** | | | | | |
| B241, B345 | 1060 | All | 8.5 | 60 | Seamless tubular products |
| B210, B234 | 1060 | All | 8 | 62 | Seamless tubular products |
| B234, B241 | 3003 | All | 14 | 96 | Seamless tubular products |
| B345 | Alclad 3000 | All | 13 | 90 | Seamless tubular products |
| **P-number 23 Aluminum-base alloys** | | | | | |
| B210, B234 | 6061 | -T4, -T6 | 24 | 165 | Seamless tubular products |
| B241, B235 | 6063 | -T5, -T6 | 17 | 117 | Seamless tubular products |

**Table 9**
**A-number classification of ferrous filler metal by chemical composition for the AR-3 level**

| Filler metal comp. no. | Reference P-number for pipe | Type of Weld deposit | AWS filler metal specification numbers and classification |
|---|---|---|---|
| A-1 | P-1 | Mild steel | A5.1, A5.2, A5.17, A5.18, A5.20 |
| A-2 | P-1 | Mild steel | A5.5 |

**Table 10**
**F-number grouping of welding rods and electrodes for carbon steel for the AR-3 level**

| AWS Specification number | F1 | F2 | F3 | F4 | F6 |
|---|---|---|---|---|---|
| A5.1 | EXX20<br>EXX24<br>EXX27<br>EXX28 | EXX12<br>EXX13<br>EXX14 | EXX10<br>EXX11 | EXX15<br>EXX16<br>EXX18 | |
| A5.2 | | | | | RXX |
| A5.5 | | | E7010-A1<br>E7011-A1 | E7018-Al | |
| A5.17 | | | | | FX-EXXX |
| A5.18 | | | | | ERXXS-X |
| A5.20 | | | | | EXXT-X |

**Table 11**
**F-number grouping of nonferrous electrodes and welding rods for the AR-3 level**

| F-number | AWS specification and classification number |
|---|---|
| | Aluminum and aluminum alloys |
| | A5.10 |
| F-21 | ER1100 |
| F-23 | ER4043<br>ER4047 |

(4) A change from multiple-pass to single-pass welding shall require requalification. For single-pass welding, the maximum thickness welded in production shall not exceed the thickness of the test pipe.

**4.3.7 Gas Shielding.** In gas metal arc, gas tungsten arc, or flux cored arc welding, a change in the composition of the gas involving one or more of the following shall require requalification:

(1) A change from one gas to another.

(2) The addition or omission of an active gas such as oxygen, hydrogen, or carbon dioxide.

(3) The addition or omission of more than 5 percent by volume of another inert gas.

(4) With gas mixtures, a 5 percent change in the volume of one of the gases.

(5) A decrease of 10 percent or more in the gas flow rate.

**4.3.8 Electrical Characteristics**

(1) A change in the welding current from AC to DC, or vice versa, shall require requalification.

(2) A change in polarity shall require requalification.

(3) For the gas metal arc process, a change in current density that results in a change from spray arc, globular arc, or pulsating arc to a short circuiting arc, or vice versa, shall require requalification.

**4.3.9 Welding Position.** The addition of any welding position other than those already qualified shall require requalification.

**4.3.10 Direction of Welding.** When the pipe is in the 5G or 6G position, a change in the direction of welding, that is from bottom to top or top to bottom, shall require requalification.

**4.3.11 Preheat Temperature.** A change of 100° F (55° C) or more in the specified preheat temperature range shall require requalification.

**4.3.12 Interpass Temperature.** A change in the minimum or maximum specified interpass temperature shall require requalification.

**4.3.13 Postheating.** With ferrous materials, a change of 100° F (55° C) or more in the specified postheating temperature shall require requalification.

A change in the maximum temperature that takes the material above a phase transformation temperature (such as the lower critical temperature for steel) shall require requalification.

**4.4 Limitations on Changes Allowable in Welding Variables Controlling the Qualification of a Welder to the AR-3 Level**

Information concerning all of the following applicable welding variables shall be recorded in the welder qualification test record. A welder qualified to one welding procedure is also qualified to weld with other qualified welding procedures, provided that the welding variables of such procedures do not vary from those of the procedure to which the welder was qualified by more than the limits for such changes listed below. If the welding variables of the production welding procedure to be used vary from those in the welder's qualification test record by more than the limits listed below, the welder shall be requalified to the new procedure.

**4.4.1 Filler Metal.** A change from one F-number to another F-number in Tables 10 or 11, or to any other filler metal, shall require requalification, except that qualification of a welder with F-numbers 1 through 4 group filler metals automatically qualifies him for that F-number group and all lower F-number groups.

For aluminum alloys, qualification with F-number 21 group filler metals shall qualify the welder to weld with F-number 23 group filler metals, and vice versa.

**4.4.2 Pipe Size and Wall Thickness.** A change from one diameter-wall thickness combination to another, as shown in Table 7, shall require requalification.

For the oxyacetylene or oxyhydrogen process, the maximum thickness qualified is the thickness of the sample weld.

**4.4.3 Backing** (for complete joint penetration groove welds only)

(1) In any welding process, the addition or omission of a preplaced consumable insert shall require requalification.

(2) The omission of a backing ring in joints welded from one side only shall require requalification.

(3) For gas tungsten arc welding, the omission of filler metal for the root pass shall require requalification.

(4) For open butt joints, the omission of a gas backing shall require requalification.

(5) The omission of a backing weld shall require requalification.

(6) The omission of a removable material as a backing device shall require requalification.

(7) For the oxyacetylene or oxyhydrogen process, the addition of backing shall require requalification.

**4.4.4 Welding Process.** A change from one welding process to any other welding process, or combination of processes, shall require requalification. Where a combination of welding processes is required, each welder must be qualified for the particular process or processes he will be required to perform in production welding.

**4.4.5 Gas Shielding.** In gas metal arc, gas tungsten arc, or flux cored arc welding, a change in the composition of the gas involving one or more of the following shall require requalification:

(1) A change from one gas to another.

(2) The addition or omission of an active gas such as oxygen, hydrogen, or carbon dioxide.

(3) The addition or omission of more than 5 percent by volume of another inert gas or carbon dioxide.

(4) With gas mixtures, a 5 percent change in the volume of one of the gases.

(5) A decrease of 10 percent or more in the gas flow rate.

**4.4.6 Electrical Characteristics**

(1) A change in the welding current from AC to DC, or vice versa, shall require requalification.

(2) A change in polarity shall require requalification.

(3) For the gas metal arc process, a change in current density that results in a change from spray arc, globular arc, or pulsating arc to a short circuiting arc, or vice versa, shall require requalification.

**4.4.7 Welding Position.** The addition of any welding position other than those already qualified shall require requalification (see paragraph 1.4).

**4.4.8 Direction of Welding.** When the pipe is in the 5G or 6G position, a change in the direction of welding, that is from bottom to top or top to bottom, shall require requalification.

# 5. Welding Procedure and Welder Qualification for Fillet Welds Only

**5.1 General Notes on Qualification for Fillet Welds**

Both welding procedures and welders qualified using sample groove welds are automatically qualified for fillet welds. However, if only fillet welds are required in production, the welding procedure and the welder may be qualified for use only on fillet welds by preparing the sample fillet weld shown in Fig. 5.2 and subjecting it to the tests and acceptance requirements in paragraph 5.3. Successful qualification using this sample fillet weld qualifies the welding procedure and the welder for all fillet sizes, material thicknesses, and diameters of pipe and tube.

The same sample fillet weld is used for qualifying both the welding procedure and the welder. The tests required and acceptance requirements are identical for both welding procedure and welder qualification.

If it becomes apparent that the welding conditions set forth in the tentative welding procedure will not produce a joint that meets the requirements, the test should be discontinued. One such indication is the need for an excessive number of repairs to produce an acceptable weld. In such a case, new conditions must be established and their validity proven by a new test weld.

A welder who prepares the sample fillet weld for the successful qualification of a welding procedure is qualified to weld with that procedure without further testing.

Paragraph 5.4 should be referred to prior to starting the procedure qualification testing to determine all of the variables that need to be controlled and documented. The limits on allowable changes in welding variables controlling the qualification of a welding procedure for fillet welds only, as listed in paragraph 5.4, should be referred to in order to determine when modification of an existing procedure requires the performance of new qualification tests.

To determine if a welder's existing qualifications will permit him to weld with a new welding procedure, refer to paragraph 5.5, "Limitations on Changes Allowable in Welding Variables Controlling the Qualification of a Welder for Fillet Welds Only."

**5.2 Preparation of Sample Fillet Welds.** Prepare a sample fillet weld of the size and type shown in Fig. 5.2. This sample fillet weld is used for qualifying both the welding procedure and the welder for all pipe sizes and wall thicknesses. Larger sample weldments are permitted for testing with the submerged arc welding process.

As an alternative, a sample fillet weld may be prepared using job size pipe and a piece of flat plate as shown below. Sample fillet welds of this configuration will qualify both welding procedures and welders for fillet welds in all pipe sizes and wall thicknesses. When using this configuration, details for sectioning the sample fillet weld and the fillet weld fracture test must be adjusted accordingly.

Orient the sample fillet weld in one of the standard test positions, and complete the weld. As set forth in paragraph 1.4, testing in the 5F positions qualifies for fillet welds in all positions; testing in the 4F position qualifies for the 2F and 4F positions; testing in the 2F position qualifies the 1F and 2F positions; and testing in the 1F position qualifies only for fillet welds in that position.

**5.3 Tests Required for Sample Fillet Welds**

**5.3.1 Visual Inspection.** Each sample weld shall be inspected visually with the unaided eye:

(1) *Prior to assembly* of the test pieces, to assure that the proper materials are being used, that the parts have been cleaned, and that the end preparation is correct.

(2) *After fit-up,* to check the joint clearances.

(3) *During welding,* to inspect for cracks, porosity, undercut, incomplete fusion, surface roughness, and poor crater conditions. Grinding may be used when necessary on starts and stops and to remove local defects from bead-to-bead, but dressing of the completed sample weld shall be limited to the removal of local defects appearing on the outer surface of the weld reinforcement.

(4) *After welding* (including any postheating), to detect any surface defects.

The following are the acceptance requirements for a sample weld to pass visual inspection:

(a) The specified base metal and filler metal shall have been used.

(b) The weld shall be free of cracks.

(c) All craters shall be filled to the full cross section of the weld and shall be free of cracks upon completion of the welding.

(d) There shall not be more than 1/8 in. (3.2 mm) difference in the lengths of the legs of the fillet weld, and the weld undercut at the edges of the legs shall not exceed 1/32 in. (0.8 mm).

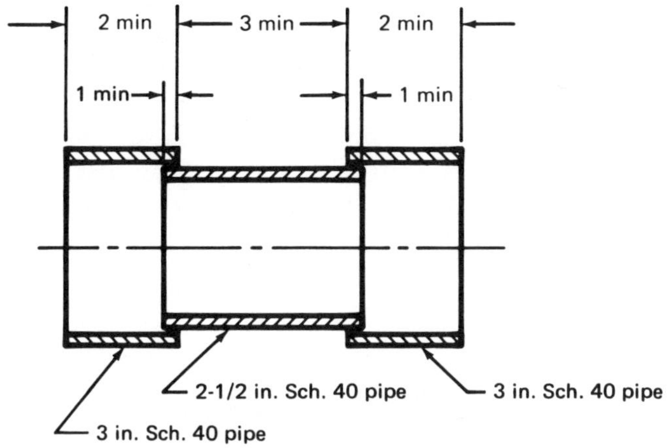


SI Equivalents

| in. | mm |
|---|---|
| 1 | 25 |
| 2 | 51 |
| 2-1/2 | 63 |
| 3 | 76 |

Note: All dimensions in inches.

**(A) Pipe assembly longitudinal section**

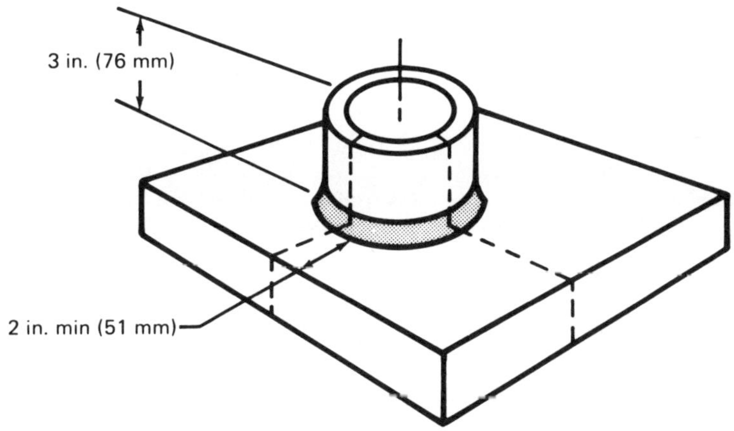


**(B) Alternate assembly**

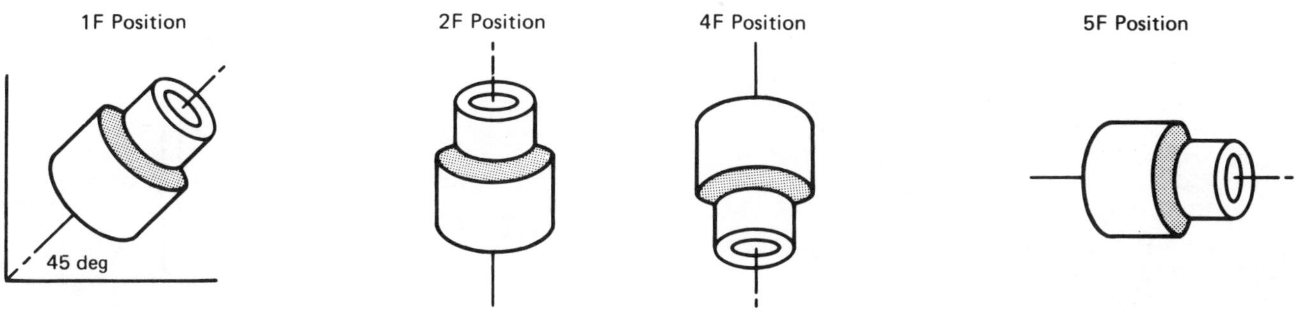


**(C) Standard test positions**

**Fig. 5.2–Preparation of sample fillet welds for welding procedure and welder qualification**

(e) Sizes of fillet welds shall be measured as shown in Fig. 5.3.1.

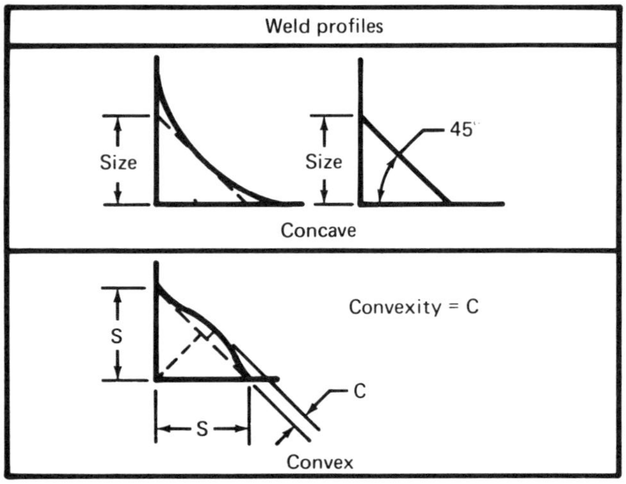


Fig. 5.3.1–Measurement of fillet welds

**5.3.2 Fillet Weld Fracture Test.** Section the sample weld for each position into four 3/4 in. (19.0 mm) wide specimens as shown in Fig. 5.3.2. Place each specimen in a vise, fracture with hammer blows in such a manner that the root of the weld is in tension, and visually inspect the fracture surfaces with the unaided eye.

In order to meet the acceptance requirements, there shall be fusion to the root of the joint, but not necessarily beyond, and no cracks. Slag and porosity visible on the fracture surfaces shall not exceed 1/8 in. (3.2 mm) measured in any direction, and the total of such indications shall not exceed 3/8 in. (9.5 mm) in any one specimen.

**5.4 Limitations on Changes Allowable in Welding Variables Controlling the Qualification of Welding Procedures for Fillet Welds Only**

Information concerning all of the following applicable welding variables shall be recorded when a welding procedure is first qualified. Changes, other than those given below, may be made without the necessity for qualifying a new welding procedure, provided that the existing welding procedure specification is revised to show these changes. The welding procedure shall be requalified and set up as a new procedure when any of the changes listed below are made in the welding procedure.

**5.4.1 Base Metal.** The grouping of materials by P-number classification is based on chemical and metallurgical characteristics. The grouping of filler metals by F-number classification is based on usability characteristics. These groupings are intended to reduce the number of welding procedure qualifications required.

(1) A change from a base metal listed under one P-number in Table 2 to material listed under another P-number, or to any other base metal, shall require requalification.

(2) A change from a similar metal joint to a dissimilar metal joint (two different P-numbers), even though welding procedures have been developed for each of the metals welded to itself, shall require requalification.

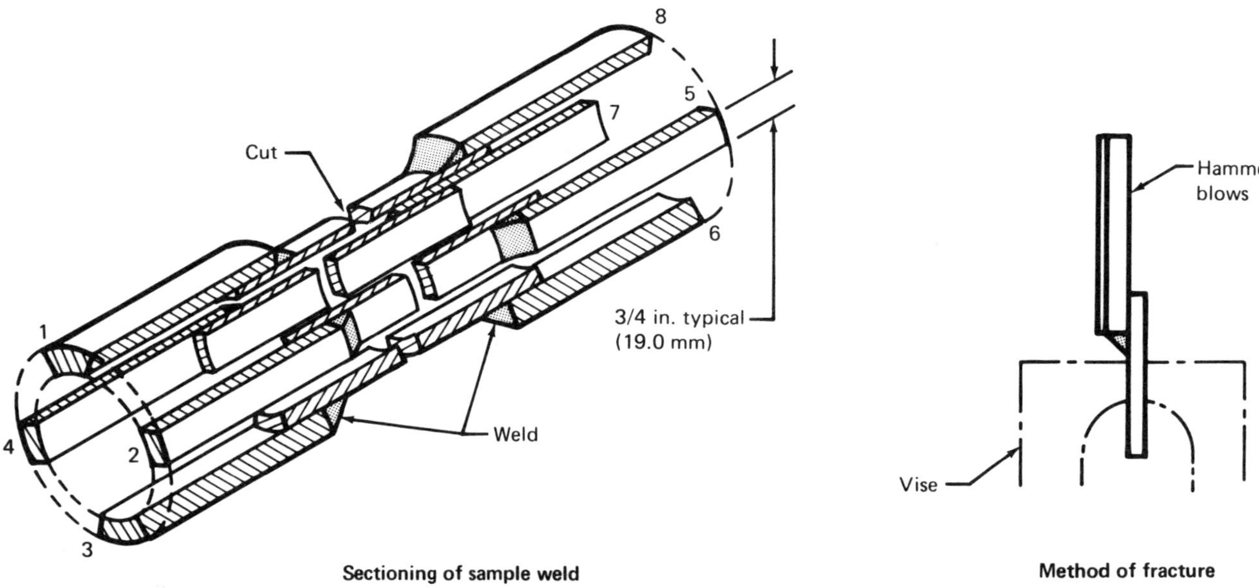


Fig. 5.3.2–Fillet weld fracture test

**5.4.2 Filler Metal**

**5.4.2.1 Composition.** Filler metal or deposited weld metal compositions other than those listed in Table 3 shall require separate qualification.

**5.4.2.2 Usability.** A change from one F-number to another F-number in Tables 4 or 5 shall require requalification.

**5.4.3 Welding Process.** A change from one welding process to another for making the entire weld or part thereof shall require requalification. Within a given process, a change from a fully automatic to semiautomatic or manual or a change from semiautomatic to manual, or vice versa, shall require requalification. Additional requirements or exceptions for various welding processes are listed below.

**5.4.3.1 Oxyacetylene or Oxyhydrogen Process.** The maximum thickness qualified using the oxyacetylene or oxyhydrogen welding process for pipe, is the thickness of the sample weld plus 20 percent.

**5.4.3.2 Gas Tungsten Arc Process**

(1) The addition or omission of filler metal shall require requalification.

(2) The omission of high frequency stabilization shall require requalification.

(3) For single-pass welding, the maximum thickness welded in production shall not exceed the thickness of the test pipe.

**5.4.3.3 Gas Metal Arc Process**

(1) For single-pass welding, the maximum thickness welded in production shall not exceed the thickness of the test pipe.

(2) A change from multipass to single-pass welding shall require requalification except for single-pass fillet welds not exceeding 1/4 in. (6.4 mm), which do not require requalification.

(3) A change in the filler metal from bare solid electrodes to metal cored electrodes, or vice versa, shall require requalification.

**5.4.3.4 Flux Cored Arc Process**

(1) For single-pass welding, the maximum thickness welded in production shall not exceed the thickness of the test pipe.

(2) A change from multipass to single-pass welding shall require requalification except for single-pass fillet welds not exceeding 1/4 in. (6.4 mm), which do not require requalification.

(3) A change in the filler metal from bare solid electrodes to metal cored electrodes, or vice versa, shall require requalification.

**5.4.3.4 Flux Cored Arc Process**

(1) For single-pass welding, the maximum thickness welded in production shall not exceed the thickness of the test pipe.

(2) A change from multipass to single-pass welding shall require requalification except for single-pass fillet welds not exceeding 1/4 in. (6.4 mm), which do not require requalification.

**5.4.3.5 Submerged Arc Process**

(1) A change in the nominal composition of flux shall require requalification.

(2) Any change that would result in the important alloying elements in the weld metal being outside the composition range given in the welding procedure shall require requalification.

(3) A change from a single arc to multiple arc, or vice versa, or a change from one type of multiple arc to another type, such as from series arc to parallel arc, shall require requalification.

(4) A change from multiple-pass to single-pass welding shall require requalification. For single-pass welding, the maximum thickness welded in production shall not exceed the thickness of the test pipe.

**5.4.4 Gas Shielding.** In gas metal arc, gas tungsten arc, or flux cored arc welding, a change in the composition of the gas involving one or more of the following shall require requalification:

(1) A change from one gas to another.

(2) The addition or omission of an active gas such as oxygen, hydrogen, or carbon dioxide.

(3) The addition or omission of more than 5 percent by volume of another inert gas.

(4) With gas mixtures, a 5 percent change in the volume of one of the gases.

(5) A decrease of 10 percent or more in the gas flow rate.

**5.4.5 Electrical Characteristics**

(1) A change in the welding current from AC to DC, or vice versa, shall require requalification.

(2) A change in polarity shall require requalification.

(3) For the gas metal arc process, a change in current density that results in a change from spray arc, globular arc, or pulsating arc to a short circuiting arc, or vice versa, shall require requalification.

**5.4.6 Welding Position.** The addition of any welding position other than those already qualified shall require requalification (see paragraph 1.4).

**5.4.7 Direction of Welding.** When the pipe is in the 5G or 6G position, a change in the direction of welding, that is from bottom to top or top to bottom, shall require requalification.

**5.4.8 Preheat Temperature.** A change of 100° F (55° C) or more in the specified preheat temperature range shall require requalification.

**5.4.9 Interpass Temperature.** A change in the minimum or maximum specified interpass temperature shall require requalification.

**5.4.10 Postheating**

(1) With ferrous materials, a change of 100° F (55° C) or more in the specified postheating temperature shall require requalification.

(2) A change in the maximum temperature that takes the material above a phase transformation temperature (such as the lower critical temperature for steel) shall require requalification.

**5.5 Limitations on Changes Allowable in Welding Variables Controlling the Qualification of a Welder for Fillet Welds Only**

Information concerning all of the following applicable welding variables shall be recorded in the welder qualification test record. A welder qualified to one welding procedure is also qualified to weld with other qualified welding procedures, provided that the welding variables of such procedures do not vary from those of the procedure to which the welder was qualified by more than the limits for such changes listed below. If the welding variables of the production welding procedure to be used vary from those in the welder's qualification test record by more than the limits listed below, the welder shall be requalified to the new procedure.

**5.5.1 Filler Metal.** A change from one F-number to another F-number in Tables 4 or 5, or to any other filler metal, shall require requalification, except that qualification of a welder with F-numbers 1 through 4 group filler metals automatically qualifies him for that F-number group and all lower F-number groups.

For aluminum alloys, qualification with F-number 21 group filler metals shall qualify the welder to weld with F-number 23 group filler metals, and vice versa.

**5.5.2 Wall Thickness.** For the oxyacetylene or oxyhydrogen process, the maximum thickness qualified is the thickness of the sample weld.

**5.5.3 Welding Process.** A change from one welding process to any other welding process, or combination of processes, shall require requalification. Where a combination of welding processes is required, each welder must be qualified for the particular process or processes he will be required to perform in production welding.

**5.5.4 Gas Shielding.** In gas metal arc, gas tungsten arc, or oxyhydrogen arc welding, a change in the composition of the gas involving one or more of the following shall require requalification:

(1) A change from one gas to another.

(2) The addition or omission of an active gas such as oxygen, hydrogen, or carbon dioxide.

(3) The addition or omission of more than 5 percent by volume of another inert gas or carbon dioxide.

(4) With gas mixtures, a 5 percent change in the volume of one of the gases.

(5) A decrease of 10 percent or more in the gas flow rate.

**5.5.5 Electrical Characteristics**

(1) A change in the welding current from AC to DC, or vice versa, shall require requalification.

(2) A change in polarity shall require requalification.

(3) For the gas metal arc process, a change in current density that results in a change from spray arc, globular arc, or pulsating arc to a short circuiting arc, or vice versa, shall require requalification.

**5.5.6 Welding Position.** The addition of any welding position other than those already qualified shall require requalification (see paragraph 1.4).

**5.5.7 Direction of Welding.** When the pipe is in the 5G or 6G position, a change in the direction of welding, that is from top to bottom or bottom to top, shall require requalification.

# Appendix A: Purpose and Mechanics of Welding Procedure and Welder Qualification

## A1.

A welding procedure is an orderly statement of the materials, detailed methods and practices, and other welding conditions that must be used to make sound welds. The welding conditions that must be established may include welding variables that can be changed at will by the welder within the limits of the welding procedure, such as welding current, travel speed, and shielding gas flow, as well as variables, such as filler metal classification of analysis, pipe wall thickness, joint design, and positions of welding that may not be changed by the welder at his discretion. The welding procedure specifies these variables so that they may be held within the proper ranges during welding to ensure the production of quality welds.

## A2.

The establishment of workable ranges or values for these welding variables is the goal of welding procedure qualification. This is accomplished by preparing sample welds using tentative values for these welding variables, evaluating these sample welds using a standardized series of tests, recording the results in a procedure qualification test record (PQR), and comparing the results with a recognized standard. This activity is called welding procedure qualification.

## A3.

Once the correct welding conditions have been established and proven by welding procedure qualifications, they are written into the welding procedure specification (WPS). The ability of welders to produce sound welds is then evaluated by having them prepare sample welds, using the qualified welding procedure, and by subjecting these welds to a standardized series of tests for welder qualification. These tests, sometimes referred to as performance qualifications, are generally less inclusive than the test series used for welding procedure qualification. The reason is that welding procedure qualification must not only determine that the tentative welding procedure will produce sound welds of correct configuration, but that the correct material properties are achieved in the weldment. Welder qualification, on the other hand, must only demonstrate that the welder has the ability to make sound welds of the correct configuration with the welding procedure to which he is being tested.

## A4.

The listing of all relevant variables in the welding procedure specification is important for several reasons. First, their values and correct ranges must be known by the welder if he is to follow the welding procedure correctly. Again, note that these include not only those welding variables that can be changed at will by the welder within the limits of the welding procedure, such as welding current, arc length, and gas flow rates, but also those variables that may not be changed by the welder at his discretion even though they may vary between welding procedures, such as pipe diameter, welding electrode classification, and joint design. By verifying that he is using the correct welding conditions specified in the welding procedure, the welder can help ensure the maintenance of quality production welds.

## A5.

Second, the presentation of all relevant welding variables in the welding procedure specification will allow that welding procedure to be compared with other proposed or existing welding procedures. By comparing the welding variables for two welding procedures, it is possible to determine if they are equivalent and, if they are not, to identify exactly those variables where differences exist.

## A6.

The committees preparing welding qualification standards have identified those welding variables that they believe have a significant effect on weld quality. These are listed in qualification standards under such names as essential variables, controlling welding parameters, or allowable limits of change. In this specification, the welding variables the Committee has included under the paragraph on controlling limitations are those in which changes will significantly alter the welding procedures in which they appear. The Committee has listed the exact amount of such changes that can be tolerated without reducing the ability of a welding procedure to produce sound welds.

## A7.

The specification requires that the exact values of the applicable welding variables be recorded in the welding procedure specification and in the procedure qualification test record. When modification of a procedure requires changes in a welding variable greater than that shown in the limits controlling the qualification, the welding procedure must be requalified to prove that the modified procedure will still produce welds of acceptable quality.

## A8.

The use of controlling limits for qualification assists in reducing the number of qualification tests that must be conducted during production to maintain weld quality. If a proposed welding procedure is equivalent to an already existing qualified procedure, qualification of the new welding procedure may be eliminated. A review of the limitations on changes in welding variables for an already qualified procedure will reveal how much modification can be made in it without requiring qualification of a new procedure.

## A9.

Since the limitations on changes in the welding variables controlling welder qualification are generally different from the limitations controlling the qualification of welding procedures, it may be possible for a welder to qualify to two or more welding procedures by qualifying with one of them, even though these welding procedures are not equivalent and each requires separate qualification. For example, a welder qualified to weld with a procedure for P-5 base metals is also qualified to weld with a procedure for P-4 base metals, even though the P-4 welding procedure must be qualified separately. A careful review of welding variables and their limitations may thereby reduce the number of welder qualification tests required.

## A10.

It must be pointed out that there are different national codes and standards that contain standardized test series for welding procedure and welder qualification. These codes and standards are directed at a variety of applications of piping systems, and the specific details of the tests and minimum acceptable test results specified may be expected to vary, depending upon the level of criticality of the systems involved. However, most of the tests and acceptance requirements for sample welds specified by these codes and standards are similar in nature.

## A11.

This specification sets forth two series of standardized tests and minimum acceptable test results for qualification of welding procedures and welders. The AR-3 level series is directed at noncritical piping systems and is discussed at the end of this Appendix. The AR-1 level test series is fairly typical of other national codes and standards for welding procedure and welder qualification, except that it is not directed at any specific application. It may be used as a guide for those developing welding requirements for new applications of piping systems. It may also assist in the understanding of the welding qualification requirements of other standards.

## A12.

For the AR-1 level of welding procedure qualification, the standardized test series requires both destructive and nondestructive tests. The required destructive tests for sample welds include tensile tests and guided-bend tests. The required nondestructive tests are visual inspection, radiography, and liquid penetrant or magnetic particle examination, if this inspection process will be used on production welds.

## A13.

The use of the tensile test, and to some extent the bend test, in the qualification of AR-1 welding procedures will determine that the welding conditions specified by the welding procedure will produce weldments with adequate material properties. For certain special applications, some other national standards also require impact testing to verify that the impact resistance of deposited weld metal will be suitable for the intended application.

## A14.

Visual, radiographic, and liquid penetrant or magnetic particle examinations are included in the required AR-1 welding procedure qualification tests to assure that it is possible, using the welding conditions specified in the welding procedure, to make sound welds meeting the minimum acceptance requirements for weld profile, penetration, and freedom from cracking, porosity, or slag entrapment.

## A15.

The standardized test series for AR-1 welder qualification are less extensive than those required for qualification of an AR-1 welding procedure. The reason for this is that welding procedure qualification must prove that the total effect of the welding conditions listed in the welding procedure specification will produce acceptable welds, both from a material properties standpoint and from a soundness and correct weld profile standpoint. Once the relevant ranges for these welding conditions have been established, the tests standardized for welder qualification must only evaluate whether the welder can make sound welds of the correct weld configuration using the qualified welding procedure. Since the evaluation of material properties is not a function of the welder qualification, the tensile test is not used for welder qualification.

## A16.

The standardized tests required for AR-1 welder qualification are divided into a destructive test series and a nondestructive test series. For AR-1 welder qualification, either of these test series may be used for evaluation of a welder's sample welds.

## A17.

The destructive test series for AR-1 welder qualification includes visual inspection and the guided-bend test for all sample welds over 2 inches (51 mm) in nominal pipe size. For small diameter (1-1/2 in. [38 mm] and under) nominal pipe sizes, sectional examination is substituted for the bend test. The destructive test series for AR-1 qualification may be utilized for all combinations of piping materials and welding processes.

## A18.

The nondestructive test series for AR-1 welder qualification is comprised of visual inspection and radiography. While both the destructive and nondestructive test series for welder qualification evaluate the soundness and correct configuration of sample welds, the nondestructive test series using radiography allows an organization using radiography in production inspection to upgrade a welder's qualification by radiographing a production weld. This will reduce the costs of welder qualification. The nondestructive test series for welder qualification may not be used for all combinations of piping materials and welding processes. Exact restrictions are given in the specification in the section on AR-1 welder qualification.

## A19.

The AR-3 qualification level is intended for applications where a high degree of weld confidence does not have to be established by performing tests on sample welds. Such cases may arise when proposed construction materials and welding procedures have been repeatedly and successfully used for similar applications in the past and when the consequences, if the system fails to contain its fluid, are minor. An example of this might be low pressure heating and ventilating piping systems, where the properties of welds in commercially available pipe are known to have given adequate service over many years. Another typical AR-3 application might be sanitary water and drain systems in which a leak would pose no threat to life or property.

## A20.

For the AR-3 level of qualification, the tests and acceptance requirements specified for sample welds are identical for both welding procedure and welder qualification. For small diameter pipe, sizes 1-1/2 in. (38 mm) and under, visual inspection and the pressure bend test are the only tests required for both welding procedure and welder qualification. For all sample welds over 2 in. (38 mm) nominal pipe size, visual inspection and macrostructure examination are the tests required for AR-3 welding procedure and welder qualification. The guided-bend test may be substituted for the macrostructure examination, but the acceptance requirements for the bend test are less stringent for the AR-3 level than for the AR-1 level.

## A21.

The required tests for the AR-3 qualification level primarily determine that the contractor's welding equipment and welders can produce sound welds meeting acceptable requirements for weld profile, freedom from cracks, and lack of gross porosity and slag entrapment.

## A22.

The specification also contains requirements for the qualification of welding procedures and welders for fillet welds only. As in the AR-3 level, the tests and acceptance requirements specified are identical for both welding procedure and welder qualification. The required tests are visual inspection and the fillet weld fracture test. These are performed to determine that the welding procedure and welder can produce fillet welds meeting the acceptance requirements for soundness and weld profile.

# Appendix B: Definitions

The nomenclature within this specification conforms to the latest edition of AWS A3.0, Welding Terms and Definitions. Other terms not included in that document but commonly used in welding piping and tubing are defined below. Clarification is also included for some terms presently included in A3.0, but considered insufficiently defined for purposes of this document.

**A-number.** A letter-number designation for grouping ferrous filler metals by chemical composition (see Tables 3 and 9).

**contractor.** In this document, the use of the term 'contractor' refers to any person or organization undertaking to qualify welding procedures and welders in accordance with this specification.

**dissimilar metal welds.** Welds between two metals of different P-numbers or groups within a P-number (see Table 2).

**F-number.** A letter-number designation for grouping ferrous and nonferrous electrodes and welding rods to reduce the number of qualification tests required (see Tables 4 and 5).

**joint penetration.** The minimum depth a groove or flange weld extends from its face into a joint, exclusive of reinforcement. Inadequate penetration of a groove weld in a butt joint is defined as penetration that is less than that specified. Therefore, in the welding of butt joints in pipe or tube, complete joint penetration is penetration to the full thickness of the joint. For fillet welds, complete penetration is considered to be fusion to the root of the joint, but not necessarily beyond.

**P-number.** A letter-number designation for grouping base metals of similar chemical and metallurgical characteristics. These groupings minimize the number of qualification tests required and indicate those cases where procedure qualification must be performed to establish the welding conditions for joining dissimilar metal combinations (see Tables 2 and 8).

**welder.** In this document, the use of the term 'welder' includes both welders and welding operators.

# Appendix C: Testing Options

## C1. Qualification with Job Size Pipe

In tubular products, both the diameter and wall thickness are variables in the welding. A concerted effort has been made to carefully select representative pipe sizes for sample welds in the qualification tables to take into account the effect of pipe or tube diameter and wall thickness. For sizes outside the ranges qualified by the sample welds, such as heavy wall small diameter tube, individual tests should be performed as appropriate.

It is also recognized that cases will arise in fabrication where specific pipe sizes will be involved that do not include the sample weld sizes. In such instances, actual job size pipe may be substituted. For pipe wall less than 3/4 in. (19.0 mm), the qualification is limited to thicknesses between one-half and two times the wall thickness of the test pipe, but not more than 3/4 in. (19.0 mm). For job size pipe having a wall thickness over 3/4 in. (19.0 mm), the qualification limits shown in the tables may be used.

## C2. Impact Tests

For certain materials used at low temperatures, impact tests should be performed. In these applications, the purchaser and the fabricator should agree upon a testing program covering the size and type of test specimen, the number of specimens to be tested, the areas from which the specimens are to be taken, and the acceptance requirements.

## C3. Bend Tests

The bend test in this specification is intended primarily as a soundness test. It is, to a lesser extent, an indication of ductility. The jig dimensions shown were selected after careful consideration of the primary purpose of the tests, the geometry of the specimens employed, and the fact that the weld metal soundness is also evaluated by mandatory radiography for the AR-1 level.

The Committee appreciates that bend radii smaller than those shown will reveal the presence of smaller defects. In those cases in which a higher sensitivity is required for the more ductile materials, the purchaser and the fabricator should come to an agreement on refinements of this test. Such refinements might include a smaller bend radius, different specimen dimensions, or more restrictive acceptance requirements.

## C4. Microstructure Examinations

For certain critical applications that involve the welding of high alloy steels, it may be appropriate to make microstructural examination of the procedure qualification sample welds. The user and the fabricator should agree upon the details of such an examination.

# Appendix D: Etchants

The surfaces of the specimens to be examined should be smoothed by filing or machining, or by polishing on metallographic papers. The etching period will vary depending upon the alloy being etched, the etchant used, and the temperature of the etchant, and should be continued until enough contrast is obtained to reveal any defects present in the sample weld. After a suitable contrast is obtained, the specimens should be thoroughly rinsed and dried with a stream of warm air. To preserve the appearance of the specimen, a thin coating of lacquer or clear acrylic may be applied to the etched surface.

Inasmuch as the etchants to be used contain acid, precaution should be used in their handling. When preparing etchants, remember that the acid parts must always be poured into the water parts of the etchant to avoid splashing the solution. All etching should be performed under a hood as a protection against fumes.

| Base metal | Etchant | Preparation | Technique |
|---|---|---|---|
| Carbon and low alloy steels | Hydrochloric acid | One part hydrochloric acid and one part water (by vol.). | Immerse the specimen in the boiling etchant. |
| | Ammonium persulfate | One part ammonium persulfate to nine parts water (by weight). | Rub the specimen with a piece of cotton saturated with the etchant at room temperature. |
| | Iodine and potassium iodide | One part of powdered iodine, two parts of powdered potassium iodide, and ten parts of water (by weight). | Brush the specimen surface with the etchant at room temperature. |
| | Nitric acid | One part of nitric acid to three parts of water (by vol.). | The specimen may be immersed in the boiling etchant or it may be applied to the specimen at room temperature with a glass dropper. |
| Aluminum and aluminum alloys | Hydrochloric and hydrofluoric acid | Hydrochloric acid, 15 ml; hydrofluoric acid, 10 ml; and water, 85 ml. | Swab or immerse the specimen in the etchant at room temperature. |
| | Keller's Etch | 1/2 cc of hydrofluoric acid, 2-1/2 cc of concentrated nitric acid, 1-1/2 cc of concentrated hydrochloric acid, and water to make 100 cc of solution. | Swab or immerse the specimen in the etchant at room temperature. |
| Copper and copper alloys | Nitric acid | Concentrated nitric acid. | Immerse the specimen in the etchant at room temperature for a few seconds. Rinse with water and repeat with a solution of one part nitric acid and one part water. For silicon-bronze alloys, it may be necessary to swab the specimen to remove $SiO_2$ deposits that may be formed. |
| Nickel | Nitric acid | Concentrated nitric acid. | Swab or immerse the specimen in the etchant at room temperature. |

| Base metal | Etchant | Preparation | Technique |
|---|---|---|---|
| Low carbon nickel, and nickel-copper (400) alloys | Lepito's Etch | 1.5 g of ammonium sulfate dissolved in 7.5 cc of water, and 2.5 g of ferric chloride dissolved in 10 cc of heated hydrochloric acid. Mix the two solutions and add 3 cc of concentrated nitric acid. | Swab or immerse the specimen in the etchant at room temperature. |
| Nickel-chromium-iron alloys (600 and 800) | Aqua Regia | One part nitric acid and two parts hydrochloric acid. | Swab or immerse the specimen in the etchant at room temperature. |
| | Lepito's Etch | See above. | See above. |
| Titanium and titanium alloys | Kroll's Etch | 1 to 3 cc of hydrofluoric acid, 2 to 6 cc of nitric acid, and water to make 100 cc of solution. | Swab or immerse the specimen in the etchant at room temperature. |
| | or (recommended alternate) | 10 cc of hydrofluoric acid, 5 cc of concentrated nitric acid, and water to make 100 cc of solution. | Swab or immerse the specimen in the etchant at room temperature. |

# Appendix E: Commercial Pipe Sizes and Wall Thicknesses

The following table lists the pipe sizes and wall thicknesses currently established as standard, or specifically:

1. The traditional standard weight, extra strong, and double extra strong pipe.
2. The pipe wall thickness schedules listed in American Standard B36.10, which are applicable to carbon steel and alloys *other than* stainless steels.
3. The pipe wall thickness schedules listed in American Standard B36.19, and ASTM Specification A409, which are applicable *only* to corrosion resistant materials. (NOTE: Schedule 10S is also available in carbon steel in sizes 12 in. and smaller.)

**ASA-B36.10 and B36.19**

| Nominal pipe size | Outside diam. | Nominal wall thickness for Sched. 5S* | Sched. 10S* | Sched. 10 | Sched. 20 | Sched. 30 | Stand-ard† | Sched. 40 | Sched. 60 | Extra strong†† | Sched. 80 | Sched. 100 | Sched. 120 | Sched. 140 | Sched. 160 | Extra strong |
|---|---|---|---|---|---|---|---|---|---|---|---|---|---|---|---|---|
| 1/8 | 0.405 | – | 0.049 | – | – | – | *0.068* | 0.068 | – | *0.095* | 0.095 | – | – | – | – | – |
| 1/4 | 0.540 | – | 0.065 | – | – | – | *0.088* | 0.088 | – | *0.119* | 0.119 | – | – | – | – | – |
| 1/8 | 0.675 | – | 0.065 | – | – | – | *0.091* | 0.091 | – | *0.126* | 0.126 | – | – | – | – | – |
| 1/2 | 0.840 | 0.065 | 0.083 | – | – | – | *0.109* | 0.109 | – | *0.147* | 0.147 | – | – | – | 0.188 | 0.294 |
| 3/4 | 1.050 | 0.065 | 0.083 | – | – | – | *0.113* | 0.113 | – | *0.154* | 0.154 | – | – | – | 0.219 | 0.308 |
| 1 | 1.315 | 0.065 | 0.109 | – | – | – | *0.133* | 0.133 | – | *0.179* | 0.179 | – | – | – | 0.250 | 0.358 |
| 1-1/4 | 1.660 | 0.065 | 0.109 | – | – | – | *0.140* | 0.140 | – | *0.191* | 0.191 | – | – | – | 0.250 | 0.382 |
| 1-1/2 | 1.900 | 0.065 | 0.109 | – | – | – | *0.145* | 0.145 | – | *0.200* | 0.200 | – | – | – | 0.281 | 0.400 |
| 2 | 2.375 | 0.065 | 0.109 | – | – | – | *0.154* | 0.154 | – | *0.218* | 0.218 | – | – | – | 0.344 | 0.436 |
| 2-1/2 | 2.875 | 0.083 | 0.120 | – | – | – | *0.203* | 0.203 | – | *0.276* | 0.276 | – | – | – | 0.375 | 0.552 |
| 3 | 3.5 | 0.083 | 0.120 | – | – | – | *0.216* | 0.216 | – | *0.300* | 0.300 | – | – | – | 0.438 | 0.600 |
| 3-1/2 | 4.0 | 0.083 | 0.120 | – | – | – | *0.226* | 0.226 | – | *0.318* | 0.318 | – | – | – | – | – |
| 4 | 4.5 | 0.083 | 0.120 | – | – | – | *0.237* | 0.237 | – | *0.337* | 0.337 | – | 0.438 | – | 0.531 | 0.674 |
| 5 | 5.563 | 0.109 | 0.134 | – | – | – | *0.258* | 0.258 | – | *0.375* | 0.375 | – | 0.500 | | 0.625 | 0.750 |
| 6 | 6.625 | 0.109 | 0.134 | – | – | – | *0.280* | 0.280 | – | *0.432* | 0.432 | – | 0.562 | – | 0.719 | 0.684 |
| 8 | 8.625 | 0.109 | 0.148 | – | 0.250 | 0.277 | *0.322* | 0.322 | 0.406 | *0.500* | 0.500 | 0.594 | 0.719 | 0.812 | 0.906 | 0.875 |
| 10 | 10.75 | 0.134 | 0.165 | – | 0.250 | 0.307 | *0.365* | 0.365 | 0.500 | *0.500* | 0.594 | 0.719 | 0.844 | 1.000 | 1.125 | 1.000 |
| 12 | 12.75 | 0.156 | 0.180 | – | 0.250 | 0.330 | *0.375* | 0.406 | 0.562 | *0.500* | 0.688 | 0.844 | 1.000 | 1.125 | 1.312 | 1.000 |
| 14 O.D. | 14.0 | 0.156 | 0.188 | 0.250 | 0.312 | 0.375 | 0.375 | 0.438 | 0.594 | 0.500 | 0.750 | 0.938 | 1.094 | 1.250 | 1.406 | – |
| 16 O.D. | 16.0 | 0.165 | 0.188 | 0.250 | 0.312 | 0.375 | 0.375 | 0.500 | 0.656 | 0.500 | 0.844 | 1.031 | 1.219 | 1.438 | 1.594 | – |
| 18 O.D. | 18.0 | 0.165 | 0.188 | 0.250 | 0.312 | 0.438 | 0.375 | 0.562 | 0.750 | 0.500 | 0.938 | 1.156 | 1.375 | 1.562 | 1.781 | – |
| 20 O.D. | 20.0 | 0.188 | 0.218 | 0.250 | 0.375 | 0.500 | 0.375 | 0.594 | 0.812 | 0.500 | 1.031 | 1.281 | 1.500 | 1.750 | 1.969 | – |

| Nominal pipe size | Outside diam. | Nominal wall thickness for | | | | | | | | | | | | | | |
|---|---|---|---|---|---|---|---|---|---|---|---|---|---|---|---|---|
| | | Sched. 5S* | Sched. 10S* | Sched. 10 | Sched. 20 | Sched. 30 | Standard† | Sched. 40 | Sched. 60 | Extra strong†† | Sched. 80 | Sched. 100 | Sched. 120 | Sched. 140 | Sched. 160 | Extra strong |
| 22 O.D. | 22.0 | 0.188 | 0.218 | 0.250 | 0.375 | 0.500 | 0.375 | – | 0.875 | 0.500 | 1.125 | 1.375 | 1.625 | 1.875 | 2.125 | – |
| 24 O.D. | 24.0 | 0.218 | 0.250 | 0.250 | 0.375 | 0.562 | 0.375 | 0.688 | 0.969 | 0.500 | 1.218 | 1.531 | 1.812 | 2.062 | 2.344 | – |
| 26 O.D. | 26.0 | – | – | 0.312 | 0.500 | – | 0.375 | – | – | 0.500 | – | – | – | – | – | – |
| 28 O.D. | 28.0 | – | – | 0.312 | 0.500 | 0.625 | 0.375 | – | – | 0.500 | – | – | – | – | – | – |
| 30 O.D. | 30.0 | 0.250 | 0.312 | 0.312 | 0.500 | 0.625 | 0.375 | – | – | 0.500 | – | – | – | – | – | – |
| 32 O.D. | 32.0 | – | – | 0.312 | 0.500 | 0.625 | 0.375 | 0.688 | – | 0.500 | – | – | – | – | – | – |
| 34 O.D. | 34.0 | – | – | 0.312 | 0.500 | 0.625 | 0.375 | 0.688 | – | 0.500 | – | – | – | – | – | – |
| 36 O.D. | 36.0 | – | – | 0.312 | 0.500 | 0.625 | 0.375 | 0.750 | – | 0.500 | – | – | – | – | – | – |
| 42 O.D. | 42.0 | – | – | – | – | – | 0.375 | – | – | 0.500 | – | – | – | – | – | – |

All dimensions are given in inches. (1 in. = 25.4 mm).

The decimal thicknesses listed for the respective pipe sizes represent their nominal or average wall dimensions. The actual thicknesses may be as much as 12.5% under the nominal thickness because of mill tolerance. Thicknesses shown in light face for Schedule 60 and heavier pipe are not currently supplied by the mills unless a certain minimum tonnage is ordered.

*Schedules 5S and 10S are available in corrosion resistant materials and Schedule 10S is also available in carbon steel.

†Thicknesses shown in italics are available also in stainless steel, under the designation Schedule 40S.

††Thicknesses shown in italics are available also in stainless steel, under the desgination Schedule 80S.

# Appendix F: Sample Forms

## Sample Form A: Typical Welding Procedure Specification (WPS)

This form may be used for all welding processes or combinations of processes. The contractor shall document each welding procedure in accordance with the format shown below, or equivalent. The appropriate information shall be obtained from data obtained during the qualification of the procedure.

**Welding Procedures Specification No:** WPS________________________ **Date:** ____________________

Title of Standard Governing Qualification: ______________________________________________

Level of Qualification: AR-1 ____________ AR-3____________

**Title:**
Welding ____________ (insert welding process or processes) of____________ (insert material) to ____________ (insert material).

**Scope:**
This procedure is applicable for ____________(insert groove and/or fillet welds) in piping systems within the range of____________(insert pipe or tube diameter) through ____________ . The range of wall thickness shall be ________ through ________ . Welding may be performed in the following positions ____________ (1G, 2G, etc.).

**Base Metal:**
The base metal shall conform to ____________________ (list the applicable P-number groupings, see Table 2). When appropriate, insert ASTM or API specification, or otherwise describe by chemical analysis. Information on backing rings or consumable inserts shall be given here.

**Filler Metal:**
The filler metal shall conform to AWS specification No. ________ from AWS specification A5.________ . This filler metal falls into F-number________ and A-number________(see Tables 3, 4, and 5). If no AWS classification exists, or if the 'G' classification is used, the manufacturer's trade name and/or weld deposit chemical analysis shall be listed. Describe any special storage and drying requirements.

**Flux for Submerged Arc:**
The manufacturer's trade name shall be listed. ______________________________________________

**Submerged Arc Weld Metal Analysis:**
List the principal alloying elements and the range permitted if not covered by an applicable AWS Filler Metal Specification.

**Shielding Gas:**
The shielding gas, or gases, shall conform to the following composition and purity: ____________________ (insert gas or gases for nozzle and backing, the proportional parts, and flow rates).

**Joint Design and Tolerances:**
Sketches shall be made to show the details such as the groove angle, root face dimensions, the root opening, type of backing, the maximum mismatch permitted, the tolerance of each dimension, and any changes in joint design based on changes in thickness.

**Preparation of Base Metal:**
The edges of parts to be joined shall be prepared by____________________________ (state whether machined, ground, oxygen cut, etc.). State cleaning methods to remove foreign materials from the mating and joining surfaces prior to welding. Joint alignment shall be maintained by____________ (state which means are used such as jig, fixtures, tack weld, etc.). If tack welds are used, describe the welding process to be used, whether they are to be removed as welding progresses, the number, and location.

**Electrical Characteristics:**
The current shall be______________(state whether direct or alternating, and if alternating, give the frequency). The base metal shall be on the______________(negative or positive when direct current is used) side of the line.

**Preheat:**
This paragraph shall describe any preheating requirements. State approved methods of heating, the temperature, and the equipment required to measure the temperature.

**Welding Technique:**
Fully describe the welding technique to be used. State whether tack welds are to be used, the angle and manipulation of the electrode holder or electrode, direction of welding, pass sequence, and precautions for gas shielding. In addition, include sketches and list pertinent welding conditions based on the production application. These shall include reasonable ranges of amperage and voltage, width of weave, gas flow, travel speed, etc., for each pass of welding. In the case of vertical welds, state the progression of each pass, whether upward or downward.

**Interpass Temperature:**
This paragraph shall describe any interpass temperature requirements, location of temperature measurement and, if applicable, the requirements for interruption of the welding and preheating prior to welding.

**Cleaning and Defects:**
State the type and extent of cleaning (grinding, chipping, oxygen gouging, etc.) required between passes and at the completion of welding to control the weld contour and remove surface defects. Include the method of preparing the underside of the weld if double-groove welds are to be used. Describe any peening operations, if applicable. Specify the appearance and maximum reinforcement of the cover pass.

**Repair:**
Provide information on repair as follows: List the method of removing any defects, the welding process and procedure for repair, the nondestructive inspection methods, and the standards used to evaluate the removal and rewelded area.

**Postheating:**
This paragraph shall describe any heat treatment after welding. Include, if applicable, the type of heating, width of heat requirements, rate of heating, time and temperature at heat, method of measuring the temperature, range of differential heating allowed, and the cooling rate.

**Marking:**
Tell how joints are to be marked or recorded to establish the identity of each welder.

**Sketches:**
Include information required in the Joint Design and Welding Technique paragraphs.

## Sample Form B: Typical Contractor's Procedure Qualification Test Record (PQR)

**Contractor:** ______________________ **Date:** ______________

The contractor shall record the applicable conditions described in this document in sufficient detail during the preparation and testing of the sample welds to assure that the welding procedure has been qualified.

Qualified to Welding Procedure No: WPS ______________ Date: ______________

PQR No: ______________ for AWS AR-1 ______________ for AWS AR-3 ______________

Welding Process(es) ______________ Type of Machine ______________

Welding Method(s):

Manual ________ Semiautomatic ________ Machine ________ Automatic ________

Material Specification ______________ to ______________

Position ______________ Pipe Diameter ______________ Wall Thickness ______________

Filler Metal Specification No./Classification No. ______________

Consumable Insert Classification No. and Type ______________

Flux Specification No., Classification No., Composition, Manufacturer's Trade Name ______________

Type of Joint ______________

Welding Technique ______________

Backing (Ring, Filler Metal, Gas, etc.) ______________

Surface Preparation ______________

Precleaning ______________

Preheat ______________

Tack Welds ______________

Method of Joint Alignment: Internal ______________ External ______________

(Clamps, etc.)

Peening ______________

Method of Arc Initiation ______________

Interpass Temperature ______________

Cleaning During Welding ______________

Back-Gouging Method ______________

Inspection Prior to Welding ______________

Inspection During Welding ______________

Inspection After Welding ______________

Postheating ______________

Space Limitations (if any) ______________

**Welding Conditions**

Shielded metal arc

| Pass no. | Electrode | | Amps | Volts | AC, DCSP, or DCRP | Stringer or weave bead | Vertical up or down |
|---|---|---|---|---|---|---|---|
| | Size | Class | | | | | |
| | | E- | | | | | |
| | | E- | | | | | |
| | | E- | | | | | |
| | | E- | | | | | |

Oxyacetylene or oxyhydrogen

| Pass no. | Filler metal | | Tip size | Gas pressure | | Forehand or backhand | Flame character |
|---|---|---|---|---|---|---|---|
| | Size | Class | | Oxygen | Fuel | | |
| | | | | | | | |
| | | | | | | | |

Gas tungsten arc

| Pass no. | Filler metal | | Consumable insert | Range of welding | | DCSP, DCRP, or AC | High-frequency | Shielding gas | | | |
|---|---|---|---|---|---|---|---|---|---|---|---|
| | | | | | | | | Composition | | Flow-cfh | |
| | Size | Class | Class/Type | Amps | Volts | | | Torch | Backing | Torch | Backing |
| | | | | | | | | | | | |
| | | | | | | | | | | | |
| | | | | | | | | | | | |
| | | | | | | | | | | | |
| | | | | | | | | | | | |

| Electrode holder | | Tungsten | | | Nozzle | | Open circuit voltage | Slope control | | Bead technique |
|---|---|---|---|---|---|---|---|---|---|---|
| Type | Angle | Class | Size | End prep. | Type | Size | | Start | Finish | |
| | | | | | | | | | | |
| | | | | | | | | | | |

Gas metal arc or flux cored arc welding

| Pass no. | Filler metal | | Range of welding | | DCSP, DCRP, or AC | Travel speed, ipm | Shielding gas | | | | Vertical up or down |
|---|---|---|---|---|---|---|---|---|---|---|---|
| | | | | | | | Composition | | Flow cfh | | |
| | Size | Class | Amps | Volts | | | Gun | Backing | Gun | Backing | |
| | | | | | | | | | | | |
| | | | | | | | | | | | |
| | | | | | | | | | | | |
| | | | | | | | | | | | |

| Electrode holder | | Slope control | | Open circuit voltage | Nozzle | | Distance from work | |
|---|---|---|---|---|---|---|---|---|
| Type | Angle | Start | Finish | | Type | Size | Nozzle | Electrode |
| | | | | | | | | |
| | | | | | | | | |

**Test Results:** ______________________________

Visual inspection results ______________________________

Liquid penetrant results ______________________________

Magnetic Particle Results ______________________________

Radiographic results

| Radiograph no. | Results | Radiograph no. | Results |
|---|---|---|---|
| | | | |
| | | | |

Tensile test results

| Specimen no. | Dimensions: Width | Dimensions: Thickness | Area | Ultimate load, lb | Tensile strength, psi | Fracture location |
|---|---|---|---|---|---|---|
| | | | | | | |
| | | | | | | |

Guided-bend tests

| Type | Specimen no. | Results | Type | Specimen no. | Results |
|---|---|---|---|---|---|
| | | | | | |
| | | | | | |
| | | | | | |

Special tests results ______________________________

______________________________

Welder's name ______________________ Identification no. ______________

Test conducted by ______________________ Test approved ______________

Test witnessed by ______________________

We certify that the statements in this record are correct and that the test welds were prepared, welded, and tested in accordance with the requirements of __________ (State title and level of standard governing qualification.)

Signed ______________________

Date ______________________ Company ______________________

**Sketch of Joint Design**

Show joint designs for each pipe size-wall thickness grouping. Include details of weld groove if used, the root face, the root opening, mismatch, and tolerances.

# Sample Form C: Typical Contractor's Welder Qualification Tests Record

This form or its equivalent shall be used to document the results of welder qualification tests and should be accompanied by the welding procedure record.

**Welder's name:** ______________________ **Identification no.** ______________

Title of standard governing qualification: ______________________

Welding procedure specification no. ______________ Test no. ______________

Welding process(es) ______________________

Base metal ______________ P-number ______________ to P-number ______________

Pipe diameter ______________ Group number ______________

Filler metal ______________ Pipe wall thickness ______________

Classifications ______________ A-number ______________

Filler metal size ______________ No. of beads or passes ______________

Approx. volts ______________ Approx. amps ______________

Pipe position: 1G______ 2G______ 5G______ 6G______ 1F______ 2F______ 4F______ 5F______

Direction of welding ______________________ (uphill, downhill)

Electrical characteristics: DCRP ______ DCSP ______ AC ______

Gas shielding: Gas(es) ______________ Flow rate ______________

Backing: Consumable insert (class and type) ______________ Backing ring ______________

Gas(es) ______________ Flow rate: ______________ Volume: ______________

**Test results:**

Visual inspection results: ______________________

______________________

Radiographic results:

| Radiograph no. | Results | Radiograph no. | Results |
|---|---|---|---|
| | | | |
| | | | |

Sectional or macrostructural examination or fillet weld fracture test results:

| Specimen no. | Results | Specimen no. | Results |
|---|---|---|---|
| | | | |
| | | | |

Guided-bend test results:

| Type | Specimen no. | Results | Type | Specimen no. | Results |
|---|---|---|---|---|---|
| Face | | | Face | | |
| Root | | | Root | | |
| Side | | | Side | | |

Pressure bend test results: ______________________

Special tests: ______________________

Test conducted by: ______________________ Date: ______________